PLANTS *and* *the*
PLAGUE

The Herbal Frontline

Marcus Harrison

First published in 2015 by Marcus Harrison
Lostwithiel, Cornwall. PL22 0ER

Copyright © Marcus Harrison, 2015

ISBN-13 978 0 9544158 9 1

CONTENTS

PREFACE

If there was perhaps but one word to conjure absolute terror in the minds of people centuries ago – notwithstanding any word associated with evil feudal landowners, bothersome monarchs, and Inquisition-like terrors – it would have been that of 'plague'. One 18th century writer summed it up thus: *"Before it are beautiful Gardens, crowded Habitations and populous Cities, behind it unfruitful Emptinefs and howling Defolation."* Quite simply the prospect of a visitation of the Pestilence, Peste, Epidemia or General Mortality as the disease was variously known, terrified the populace who largely lived in squalor or poor conditions, while many were ill fed and as a result were probably in relatively poor health generally; circumstances that delivered *victims in waiting* for the next pestilential visitation, be it the bubonic, pneumonic or other form.

In this book I want to examine the response of herbal medicine to this devastating disease or, rather, some of the plants used in remedies concocted to cure the symptoms and ailment. To be frank I do not 'do' herbal medicines, coming from a background of teaching about edible wild plants, and ethnobotanical research, but some of the plants that are encountered in my wild food world also find their way into the remedies against plague. In my view many of these remedies are laughable, or at least they would be if they weren't being peddled as cures against what was one of the most devastating diseases of past centuries. Included are a number of herbal plants that do not fall within my *edible* category whatsoever, yet they were obviously esteemed for their active constituents, real or imaginary.

To comprehend the way in which the plants were used it is really necessary to understand the medical context of the plague for those who prescribed herbal remedies in the past, so the content of this book covers two main areas – the contemporary view or understanding of the disease through the ages, and then the plants themselves. There are numerous more plants (and other exotic substances too) that were incorporated into remedies but my shortlist here really focuses on a number of common and wild plants rather than obscure species. I hope you will find this journey into the past as interesting as it has been for me to do the research.

As I finalize the words of this book at the start of 2015, there is the serious Ebola virus outbreak across West Africa which may, perhaps, readjust the way countries think about movement of their citizens between continents. Ebola, at this stage in the 21st century, and currently regarded as a disease without a cure, has parallels with the 14th century plague pandemic. Ebola has not yet developed a *pneumonic* form but if that were to happen then possibly a rethink on all 21st century urban living might be necessary. But then that is down to the boffins that know how best to keep us from harm.

Lastly, I would like to extend a big thank you to the staff of the rare book and manuscripts departments at the British Library, for their courteous and helpful guidance through the vast repository of knowledge held by our National Library and which has been a backbone for the researches that made this book possible.

'*Dance of Death*' by Michael Wolgemut from The Nuremberg Chronicle [1493].

INTRODUCTION

In Europe today we have largely forgotten about the plague which, in any case, can now be treated with antibiotics, although pockets of the disease do appear from time to time around the world, even now. The nearest that most of us get to the Black Death, which wiped out 20 to 25 percent of the English population in the 14th century, or to the Great Plague of London in the mid-17th century, is to be reminded of those tumultuous events by a television programme or book. Interestingly, there were even small plague outbreaks during the autumn of 1900 in Glasgow, in Essex during 1909, and also Somerset around the same time.

Although mortality figures for the great London epidemic were sometimes well documented they are not precise, and exact figures for the Black Death will *never* be known, though some 14th century eyewitnesses on continental Europe talk of individual towns or provinces where 40 to 60 percent of the population died. Extend that to today's highly interconnected global economy and one wonders how we would cope; though our more hygienic and health-conscious existence, at least in the industrialized western nations, probably tends to make a serious reappearance of a plague pandemic less likely. Well hopefully. Conversely, the close proximity of vast urban populations now living cheek by jowl may counterbalance those advances in health and hygiene, with plague likely to spread quite rapidly through the population in such circumstances. In our defence of an outbreak, however, we do have public medical infrastructures in place, available drugs, and popular and social media able to broadcast instructions and information quickly, so hopefully any outbreak could be rapidly quarantined and bottled up. That said, the resiliance of our finely-tuned infrastructure under the weight of a plague epidemic could quickly be found wanting.

For our ancestors there were no such fallbacks and plague seems to have been an ever-present, dark, threat in the background of everyone's lives. The sense of foreboding and helplessness found in many of the old texts that give accounts of the plague is exemplified by William Bullein, who comments in March 1564 that it was a disease '... *fearyng no kyng, queene, lorde, ladie, bond, or slaue, but rather maketh all creatures alike to him.*'

Such was the doom-laden view of the disease among populations, in addition to the religious strictures of past centuries, that individuals struck by plague often gave themselves up as lost with no hope of recovery from the very outset, sometimes believing that their ailment was the punishment of God. What they were fighting was a bacterium, not the wrath of God. You get a sense of the help-lessness from the Parisian physician Simon de Covino who witnessed the terrible events of the 14th century in France: "*The pestilence stamped itself upon the entire population. Faces became pale, and the doom which threatened the people was marked upon their foreheads. It was only necessary to look into the countenances of men and women to read there recorded the blow which was about to fall; a marked pallor announced the approach of the enemy, and before the fatal day the sentence of death was written unmistakably on the face of the victims. No climate appeared to have any effect upon the strange malady. It appeared to be stayed neither by heat nor cold. High and healthy situations were as much subject to it as damp and low places. It spread during the colder season of winter as rapidly as in the heat of the summer months."*

Nor did plague epidemics bring out the best in human nature. Certainly in the mid-14th century there are a number of references as to how people reacted socially to the situation. Guy De Chauliac, one of the most important medical sources from the period of the Black Death in France, tells us: *"A father did not visit his son, nor the son his father. Charity was dead. The mortality was so great that it left hardly a fourth part of the population. Even the doctors did not dare to visit the sick from fear of infection, and when they did visit them they attempted nothing to heal them, and thus almost all those who were taken ill died, except towards the end of the epidemic, when some few recovered."*

In his *Dialogue* [1578], Bullein's reflection on the social aspect is somewhat different, mentioning that there could be sudden outbreaks of devotion and generosity: *"And when thei are touched by the fearfull stroke of the Pestilence of their nexte neighbour, or els in their owne familie, then thei vse Medicines, flie the Aire, &c. Which indeede are verie good meanes, and not against Gods woorde so to doe; then other some falleth into sodaine devotion, in giuyng almose to the poore and needie, whiche before haue doen nothing els but oppressed theim and haue dooen them wrong..."*

The written eyewitness accounts of the main 14th century European plague pandemic were hardly reassuring reading for the generations that followed, and these texts found their way into both the writings and mindset of subsequent medical literature, and plague myths, for nearly four hundred years. Indeed, as you read through this book you will see how little knowledge and understanding there was about the plague until well into the 19th century. Every time a new plague outbreak appeared, or a new medical book was written on the subject, the same old scratched record was played, so perpetuating the ignorance and superstitions about the disease that was so deadly to our ancestors. There were certainly physicians who tried new treatment ideas, while others, such as Thomas Sydenham in the 17th century, attempted to understand the disease through methodical analysis of the facts, but medics were largely at a loss to understand the pathology of the disease they faced and so medical thinking on the subject remained fixed in a groove for centuries.

Popular literature, too, had a role in disseminating and maintaining the plague terror factor. In his Introduction to *Decameron* Boccaccio refers to the 1348 epidemic in Florence (whose population had been weakened by famine the previous year), describing to his audience that such was the contagious nature of the plague that by simply conversing with a victim you could be infected: *"The disease was communicated by the sick to those in health and seemed daily to gain head and increase in violence, just as fire will do by casting fresh fuel on it. The contagion was communicated not only by conversation with those sick, but also by approaching them too closely, or even by merely handling their clothes or anything they had previously touched."*

Perhaps the most widely spread picture that Boccaccio painted for future generations revolves round his story of two pigs, transmuted to canines in other, later, folklore-ish accounts of the plague: *"Such, I say, was the deadly character of the pestilential matter, that it passed the infection not only from man to man; but, what is more wonderful, and has been often proved, anything belonging to those sick with the disease, if touched by any other creature, would certainly affect and even kill it in a short space of time. One instance of this kind I took special note of, namely, the rags of a poor man just dead having been thrown into the street, two hogs came by at the time and began to root amongst them, shaking them in their jaws. In less than an hour they fell down and died on the spot."* That terrifying image of the pigs dying on the spot, as evidence of the virulency of the plague, is repeated time and again in learned medical works on the plague for centuries afterwards, and no doubt circulated in popular oral tradition too.

For example, the thread of Boccaccio's pigs and victim's clothes surface in Andrew Boorde's *Compendyous Regiment or Dyetry of Health* [1542] published about two hundred years after the *Decameron* references: *"A man cannot be to ware, nor can not kepe hym selfe to well from this syckenes, for it is so vehement and so parlouse, that the syckenes is taken with the sauour of a mans clothes the whiche hath vysyted the infectious howse, for the infection wyl lye and hange longe in clothes. And I haue knowen that whan the strawe & russhes hath ben cast out of a howse infectyd, the hogges the whiche dyd lye in it, dyed of the pestylence..."* Did Boorde actually witness pigs dying or was he simply re-packaging Boccaccio's reference for dramatic effect?

Three hundred years on from Boorde the story regarding the peasant's clothes is again trotted out in *An Historical Account of the Plague and other Pestilential Distempers* [1832], as if the scene was just freshly observed: *"... which two hogs seized with their teeth, and after they had shaken them a little, they wheeled about once or twice and fell down dead, as if they had been poisoned."*

Simon de Covino's contemporary 14th century accounts were also less than reassuring on the contagious nature of plague: *"It has been proved that when it once entered a house scarcely one of those who dwelt in it escaped... It happened also that priests, those sacred physicians of souls, were seized by the plague whilst administering spiritual aid; and often by a single touch, or a single breath of the plague-stricken, they perished even before the sick person they had come to assist."* [Opuscule relatif à la peste de 1348]

While 14th century physicians, and those in the 17th century London epidemic, were only too aware of the deadly nature of the plague, and that it was readily transmissible, they were at a loss when it came to curing the illness. Boccaccio again: *"To cure the malady neither medical knowledge nor the power of drugs was of any avail, whether because the disease was in its own nature mortal, or that the physicians could form no just idea of the cause, nor consequently ground a true method of cure; of those attacked few or none escaped, but they generally died the third day from the first appearance of the symptoms, without a fever or other form of illness manifesting itself."*

Once a victim was infected with bubonic plague (the form of disease relevant to the Black Death and London epidemics) then it was a matter of luck, or perhaps *destiny* for those who believed in such things, as to whether they would survive to reach the other side of a very dark tunnel. Recorded cases in the Black Death epidemic show that the key symptoms were serious inflammation of the throat and lungs, violent chest pains, vomiting and spitting blood, and foetid breath or odour coming from the bodies of the sick. There were also the swellings – buboes, sometimes referred to as *pushes* in the old days – that give bubonic plague its name.

Reading through contemporary medical accounts a victim's body sounds to have been in a real mess physiologically as this very unpleasant disease ran its course, either killing the individual or giving them a close brush with death. The Byzantine Emperor John Cantacuzenus, who was an eyewitness to the Black Death in the east Mediterranean region, wrote an account of his observations. He tells us how the strong and weak were equally struck down and that the medics realized they were powerless, regarding the disease as virtually incurable. The course of the malady was not the same in all cases; some people died suddenly, others a few hours after the first symptoms appeared, some within a day. Cases where the victim lingered for two or three days usually began with a violent fever. Where the disease affected the brain, the patient lost the use of speech, uttering only inarticulate sounds as if the tongue was paralysed, and died soon afterwards. Patients in this condition often became insensible to what was taking place about them, and appeared to sink into a deep sleep.

In another form the disease attacked the lungs and respiratory tract first, causing serious inflammation and sharp pains in the chest accompanied by the vomiting of blood and foetid breath. Parched by fever the patient's throat and tongue became black and congested with blood, and little seemed to quench the sufferer's thirst. Cantacuzenus also describes the restlessness and sleepless-ness of some victims, and the livid purple and black plague spots frequently broke out on the skin, but also notes that those who managed to survive a first attack largely escaped further infection. He further comments that curative remedies which worked for one person were almost poisonous to another.

Whenever the plague visited any country's population it seriously disrupted economic and social life. The London plague epidemic of 1665-6 caused serious disruption to trade and commerce, as did those in Marseilles [1720] and Moscow [1771] in the 18th century. None of these, however, was as devastating to the fabric of society as the Black Death; the economic impact of this significantly altering the way of life in many countries, including Britain. As a result of wiping out whole families and villages there were fewer people available to simply make the economy function or work the land, and as a consequence the price of labour rose as survivors were able to pick and choose their employers and demand higher wages, while the vacant houses of the gentry and landed classes found new occupants taking advantage of the turbulent times. The social and economic impacts of the plague are not within the scope of this book, but if you are interested in the history of the Black Death and its social consequences then a good starting point is the highly detailed *History of Bubonic Plague in the British Isles* by J. F. D. Shrewsbury, published in 1970 by Cambridge University Press.

Plague epidemics, it could be said, were the ultimate stress test for any State in the past, and a real sense of the demise and collapse of society can be gleaned from some of the 14th century eyewitnesses whose personal accounts could almost come straight from the scenario of an apocalypse movie. Gilles Li Muisis [c.1272-1352], an Abbot in the French town of Tournay, writes: "*It is impossible to credit the mortality throughout the whole country. Travellers, merchants, pilgrims, and others who have passed through it declare that they have found cattle wandering without herdsmen in fields, towns, and waste lands; that they have seen barns and wine-cellars standing wide open, houses empty, and few people to be found anywhere. So much so that in many towns, cities and villages, where there had been before 20,000 people, scarcely 2,000 are left; and in many cities and country places, where there had been 1,500 people, hardly 100 remain. And in many regions, both lands and fields are lying uncultivated. I have heard these things from a certain knight well skilled in the law, who was one of the members of the Paris Parliament. He was sent, together with a certain Bishop, by Philip, the most illustrious King of France, to the King of Aragon, and on his return journey passed through Avignon. Both there and in Paris, as he told me, he was informed of the foresaid things by many people worthy of credit.*"

The same sense of destruction is found in the rather flowery words of Sir Richard Blackmore's *Discourse upon the Plague*, published over three hundred years later in 1722, but the language conveys the public anxiety about the plague, as the population: "... *convey their Infection from Houfe to Houfe, and from Town to Town,* [they] *lay defolate the Country, and depopulate great Cities. This high Venom advances with refiftlefs Fury, and fweeps away the Inhabitants of the Land, like a fwelling Inundation: It fills all Places with Slaughter, triumphs in Deftruction, and is fo far from being enfeebled while it is ventilated by the Winds, and diffus'd through the Air, that it acquires more Strength by converting into its own Nature the Exhalations and Vapours that it meets with in its Way; and by this Means, like a fpreading Conflagration, it gains greater Force as it goes on, and grows more contagious."* The notion of a plague epidemic gathering strength by assimilating other airborne matter is fairly typical of many old writers and medical thinking.

The two editions of Blackmore's work that circulated around this period were in response to a plague epidemic that hit Marseilles during 1720, and in which an estimated 100,000 people died. From across the English Channel Britain's medics and Establishment monitored events warily from a distance, knowing only too well how readily the disease could spread in their direction, as it had done in the 14th century. What was most certainly different about the response of Britain's intelligentsia of the period was that of discussing plans and organization should trouble strike. Rules, regulations and dictats concerning plague epidemics and plague victims had certainly appeared over the centuries since the Black Death and Great Plague of London to be sure, but what we now see are learned men, men of science as it were, stepping forward with ideas and possible solutions as the prospect of plague focused the mind.

For example, around October 1721 Sir John Colbatch, of the College of Physicians, published a twenty page pamphlet for distribution among the movers and shakers of the time, though it went by the rather verbose title of *A Scheme for Proper Methods to be taken, should it please God to visit us with the Plague.* Colbatch's text only deals with the medical response to a London plague scenario but his plans also included the provisioning of food and blankets, as well as the stationing of militia in case of civil unrest. Given that the date is 1721 his plans would probably not seem out of place even in our own time:

I. *That the Citys of London and Weftminfter be divided into diftinct Diftricts.*

II. *That in each Diftrict there one or more Phyficians to have the immediate Care and Super-intendency of it; and that fome young Phyficians be taken in to the affiftance of the Superiors, if they find it neceffary.*

III. *That the Diftrict appointed to each Phyfician, may be as near his own Habitation as poffible; and that the Surgeons and the Apothecarys be within the Limits of it.*

IV. *That there be fix Surgeons, and fix Apothecary's in each, to be chofen by the Phyfician or Phyficians of the Diftrict, and to be under their immediate Direction.*

V.   *That ſuch Medicines as ſhall be thought neceſſary, be immediately prepared, and a convenient Room or Rooms taken for that uſe.*

VI.   *That Women for Nurſes and other uſes be had in readineſs.*

VII.   *That a Clerk and an Aſſiſtant or two, be appointed to receive and iſſue out Money, and to keep the Accounts.*

VIII.   *That the Phyſicians of each Diſtrict do pitch on a Piece of Ground to erect an Infirmary upon, ſhould there be occaſion for any; and that the Owners of ſuch Ground be obliged to part with it, by Act of Parliament, for ſo long time as it ſhall be wanted, for a valuable conſideration.*

At a personal level citizens were often reminded in medical texts of the need to keep both themselves and their households clean. Among a number of preventative household actions to be undertaken Thomas Willis advises his readers, in a *Plain and Easie Method,* that for: "... *the Air* [that] *we breathe in may be wholeſome, all Things that may advance or add to the corruption of it, ſhould diligently be removed; our Houſes, and Streets kept clean; all Filth, and whatever may cauſe noiſom ſmells, be taken away; and amongſt other things, the ſmell of Sope Suds, and Lye, in the waſhing of Clothes, be avoided; this, Experience has taught to be very Dangerous...*" Quite what the connection between 'the smell of sope suds' and danger was I am uncertain, but the concept pops up from time to time in older plague-related texts.

When the Black Death appeared in the 14th century Europe was largely an agrarian and agricultural society with most trades and jobs being land-related. With the exception of urban clusters the relatively small population was spread out over a wide area and for anyone travelling through this world it would have felt sparsely populated even before the plague struck.

KEY FIGURES, SOURCES and TYPOGRAPHY

Although many medical books were written in the past about the subject of the plague, there is a core of writers who were key players in understanding and recording the disease. Their works were frequently referred to in the catalogue of plague treatises of other writers and below is an outline of some key individuals.

Guy de Chauliac (ca. 1300 – 1368)
From the perspective of the Black Death this 14th century French physician-surgeon is important because he was not only present during the plague outbreaks of 1348 and 1360, but he also took time to document the symptoms and progress of patients under his care. He made distinctions between bubonic and pneumonic forms of the disease, and as part of his treatment regime advocated air purification, blood-letting, and a healthy diet. Although his keynote work 'Chirurgia', published in 1363, centered around general surgery and medicine he included a section on the plague. De Chauliac would become papal physician to Pope Clement VI (1342–1352) in Avignon, and then to Popes Innocent VI and Urban V, which no doubt gave him access to medical manuscripts in papal libraries.

Ambroise Paré (c.1510 – 1590)
Paré is perhaps better known for his pioneering advances in the field of military surgery, serving as a barber-surgeon under various French kings. Observant and open-minded Paré advocated the use of more benign treatment for wounds, such as his famous salve of egg yolk, turpentine and rose oil as an alternative to cautery to arrest bleeding. This enlightened philosophy extended to battle-field amputations too, where he introduced ligatures to stem arterial blood loss rather than cauterize the vessels, and he also explored limb prosthetics for the post-operative casualties. Paré wrote quite extensively, including a book on treating gunshot wounds [1545] and a 'Traicté de la pesté...' [1568] that also dealt with petite variole (smallpox) and rougeole (measles). Paré's observations on the subject of the plague were probably as equally thoughtful and probing as those of his on surgery and for this reason he has been included as a reference; Thomas Johnson's English translation of Paré's works [1649], which incorporates Paré's 1568 plague treatise, being the particular source.

William Bullein (15?? – 1576)
Bullein appears to have had a somewhat colourful life; at one stage being ship-wrecked on a sea voyage and also spending time in a debtor's prison. He wrote several books including 'Gouernement of Healthe' around 1558-9, also 'Bulwarke of defence against all Sickness...' [1562] and then his 'Dialogue against the Fever Pestilence' [1564] in which he discusses health regimes to fight against pestilential fever. To a certain extent 'Dialogue' very much reflects the thinking on plague and pestilence at the time rather than advancing medical knowledge. Rather, the text is written for edification of the lay-person, not medics, and is therefore

interesting in a socio-medical context. Later editions of this surfaced 1573 and 1578 with some modifications to the text.

Isbrand van Diemerbroek (1609 – 1674)
This Dutch physician was working in Nijmegen during several plague outbreaks between 1635 and 1637 and catalogued the case histories of around one hundred of his patients. His experiences on treatment of the disease were published in 1646 in *De Peste* which was frequently referred to for several generations of physicians, with an English translation of the original Latin work published in 1722. Unlike many of his contemporaries he rejected bloodletting as a form of treatment.

Thomas Willis (1621 – 1675)
Willis' specialisms were neuroanatomy and neurology (rather than as a plague doctor), and he is regarded as one of the fathers of neuroscience, having named many parts of the human brain and advancing a medical discipline which had developed little beyond the Galenical-influenced work of da Vinci and Vesalius. He worked between London and Oxford and lived through the key mid-17th century plague outbreaks. Around 1666 Willis collated various remedies that he had used against the disease and these were published posthumously in 1691 as 'A Plain and Easie Method of Preserving Those That are Well From the Plague' by J. Hemming, an acquaintance. The remedies within the text very much represent a working doctor's medicine chest against the plague and are referred to many times in the century or so following Willis' death.

Thomas Sydenham (1624 – 1689)
Sydenham is regarded by some as the Father of English Medicine, and lived through the turbulent times of the English Civil War and Great Plague of London, although he did not remain in London and witness the epidemic. Less dogmatic than many physicians of the period in his approach to medicine, medical ethics, treatment of patients, and pharmacy, he wrote extensively on medicine, maintained personal journals, and properly identified St. Vitus Dance as a disease. His treatment regimes and remedies were extensively quoted by physicians of the period and beyond, and he was one of the physicians who tried to solve the mysteries behind different types of fever.

Nathaniel Hodges (1629 – 1688)
When the Great Plague hit London around Christmas 1664 Hodges was one of those physicians to notice the first suspected cases of the disease. By autumn the following year the London epidemic was present in earnest and Hodges remained in the city treating victims. His own self-preservation from the plague – at least in his view – was down to eating meat with pickles and relishes, drinking large quantities of sack, and anti-pestilential preservatives, though Hodges suspected that he personally brushed with the disease a couple of times without ever developing serious consequences. In 1672 he published an account of the plague, *Loimologia*,

with excellent observations on the nature of the plague and its treatment. Hodges' work was widely respected for many generations.

William Boghurst (1631 – 1685)

This little known apothecary-physician left behind one of the best medical accounts of the Great Plague of London written from his personal experience in and around St. Giles in the Field where he kept a shop. Boghurst's manuscript (BL Sloane MS 349) was never published in his lifetime and is very much a personal and detailed grass-roots description of events, and suggests the plague started at the end of 1664 and that Boghurst had knowledge of cases of plague in the parishes of St. Giles, St. Clement's, St. Paul's, Covent Garden, and St. Martin's over the previous three or four years. The accepted story through the centuries has been that the Great Plague came suddenly to London through contaminated bales of cotton imported into the city, with the first deaths appearing in Westminster. Boghurst describes how plague gradually moved from the West End through Holborn and Strand to the City then onward to the eastern suburbs. Entitled *Loimographia*, his work was finally published for all to read in the late 19th century.

Richard Mead (1673 – 1754)

English physician Mead was ultimately to be appointed as physician to George II in 1727 but from the very start of the 18th century had been quite prominent in researching and writing about medicine – producing a treatise on poisons in 1702 and a paper on scabies. In 1720 he published the influential '*Short Discourse concerning Pestilential Contagion, and the Method to be used to prevent it,*' which became an important English language source of information on plague.

Two other key figures, among a host of others, that pop up in some of the references are those of the *ancient* physicians Avicenna and Galen. Both of these individuals were among the key influencers – rightly or wrongly – of early Western medical philosophy with physicians often sticking doggedly to their theories. Indeed, the strict adherence to Galenic medical doctrine was one of the core reasons Western medicine remained in a developmental rut for centuries.

Avicenna

Avicenna is the Latinized name of a Persian polymath who lived around the period of the 10th and 11th century AD divide. He wrote two highly influential medical works '*The Book of Healing*' and '*The Canon of Medicine*', among many others. These two texts had an enormous influence on medieval medicine with Avicenna often quoted as a source for remedies and medical philosophy by writers that followed him.

Galen

Living around the period of the 1st and 2nd century AD divide, this Greek speaking Roman physician-surgeon was, after Hippocrates, one of the most influential medical researchers of the ancient world. Indeed, his theories dominated Western medical thought for over a thousand years. Galen championed the Hippocratic notion of bodily *Humours*: imbalances in blood (sanguine), yellow bile (choleric), black bile (melancholic), and phlegm (phlegmatic) influencing human temperament. The influences of this theory remained fashionable in Western medical debate and medical literature until the late 18th century. And if it had not been for William Harvey's challenge to entrenched Galenic thought on the circulatory system in the early 17th century, physicians would have been working under the assumption that there were two separate one-way systems for venous and arterial blood, rather than a direct connection between heart activity and blood circulation.

Original Typography

Many texts from the past, both printed or handwritten, relayed their words in phonetic form as opposed to the mostly standardized form we use today, so where possible quoted excerpts in this book try to reflect the style and feel of the original as closely as feasible without alteration. Grammatical structures and punctuation were very different in the past and if you are not familiar with old texts you may initially find that you need to read through extracts a couple of times to follow the author's thread of thinking. However, once you become familiar the old grammatical formats then your reading of the texts will become more fluid.

Where there appears to be an obvious, almost intrusive mistake due to an archaic word form in a quoted section of text, or what looks like an error, the word is often accompanied by the Latin adverb [sic] – short for *sic erat scriptum*, 'thus was it written' – to show that it is not a translation error. Corrective text inserts, to help make sense of some phrases, are inserted in square brackets too.

Also within the texts is use of the 'long S', which was prevalent in texts up to the start of the 19th century and is represented by 'ʃ'. In some texts you will also find 'u' and 'v' used interchangeably so, for example, 'approuued' is 'approved'.

Another aspect of very old texts is the appearance of accented letters to replace missing consonants. The character ē indicates a missing 'n' following the 'e'; so the printed English words mē and comē are, in fact, 'men' and an old spelling of 'common', and in French vī would be 'vin', wine. Another, similar, shorthand which occurs in Latin texts, but also archaic English and French, is where a small character that looks like the number 9 is placed beside a letter. This character represents the word ending 'us', so Silvi⁹ is the Latin word *Silvius*, while pl⁹ in French is *plus*. You will find this typographic convention used in some of the earlier 14th and 15th century texts included here.

In this book passages from old texts are extensively featured in italics. Stylistically, many original texts often used italics extensively for emphasis and these will appear as *regular* text characters when reading through the quotes.

APOTHECARIES & OLD MEDICAL TERMS

Prior to the advent of official pharmacies and dispensing chemists in the 19th century it was mainly the neighbourhood or village *apothecary* who made up and doled out medicines prescribed by physicians or on their own account. Physicians often worked closely with 'chosen' apothecaries, undoubtedly to the financial detriment of patients if the contemporary written accounts are to be believed. Physicians could collect the equivalent of a finder's fee from an apothecary for delivering custom, while apothecaries themselves were not past peddling completely useless, junk, remedies. For many centuries physicians and apothecaries were in fact at each other's throats; physicians wishing to make up their own medicines and so reap any financial gain for themselves personally, while the apothecaries viewed the making of medicines as open to others too.

Over the centuries the split between physician and potion-making began to diverge, with the emergence of apothecaries as a distinct entity in the medical world much to the irritation of physicians. However, unscrupulous and even useless apothecaries were not uncommon and so their trade was eventually legislated upon in Britain in 1815, and would-be apothecaries required to take exams on the subject.

According to the 1815 Apothecaries Act: "...*it shall not be lawful for any person or persons (except persons already in practice as such) to practise as an apothecary in any part of England or Wales, unless he or they shall have been examined by the Court of Examiners, or the major part of them, and have received a certificate of his or their being duly qualified to practise as such from the said Court of Examiners; who are authorised and required to examine all person and persons applying to them, for the purpose of ascertaining the skill and abilities of such person or persons in the science and practice of medicine, and his or their fitness and qualification to practise as an apothecary.*"

The Act applied also apothecaries' assistants who had to be trained too, while the legislation stipulated that unless an apothecary was licenced then they could not recover debts through the Courts: "...*no apothecary shall be allowed to recover any charges claimed by him in any Court of Law, unless such apothecary shall prove on the trial that he was in practice as an apothecary prior to, or on the said 1st day of August, 1815, or that he has obtained a certificate to practise as an apothecary.*"

By the middle of the 19th century the role of the apothecary had developed into a professional, regulated, medical 'trade'. That said, a new breed of *druggists*, operating from shops, had stepped into the apothecaries' old shoes and there were calls for this emerging sector of the medical trade to be regulated too. There were individuals who sold snake oil, to be sure, but the new generation of pharmacists and pharmacologists took pride in their work.

The ire and contempt with which physicians and apothecaries were sometimes held can be seen in the following scathing extract from a subscription magazine called *The Family Oracle of Health* and dated 1824: "Apothecaries' draughts, *indeed, usually sent in to the number of three or four a-day, are the death of more persons in England, than either damp, cold, or unscientific eating... If your apothecary (since you must have one), sends you in two, or three, or four draughts a-day with a proportional number of pills and powders, you may to a certainty pronounce him to be a jackal... No disease known to us can possibly require such hourly drenching, and cramming the stomach with poisonous drugs, as is constantly practised. One effective dose in very violent diseases, repeated two or three times according to necessity, is enough of drugging in all conscience. More than this has a greater chance to do harm than good; but your jackal makes no distinction; you must have your regular number of draughts, whatever be your disease. But take our serious advice—always measure the honour and skill of your apothecary by the fewness of his drugs; and his meanness and knavery by their hourly number.*"

The other aspect of medicine that the current reader will need to understand is the medical language and terminology used through the centuries. Although some of the words are somewhat self-evident others are a complete mystery to the medical lay-person. Below are explanations of some of the terms used in relation to old text extracts in this book.

Alexipharmic – A medicine used to neutralize a poison. The word is sometimes used interchangeably with sudorific.

Abstergent – A medication that cleanses or clears away foulness.

Ague – Sometimes called intermittent fever in the past, ague is not a precise disease but a general febrile condition of alternating fever, sweating, chills and shivering. During intermissions the patient may be free from fever but is often weak. Use of the term varied, but the best candidate for something like this would be malarial fits, and herbal anti-fever potions were often the same or similar to some of those for controlling pestilential fevers.

Alterative – A medicine which re-establishes the healthy functions of the body without producing any evacuation by way of perspiration, purging or vomiting.

Apostem – An impostume, or abcess.

Apozem – A form of decoction (see below), especially one to which other medicinal ingredients are admixed.

Balneo, B.M. – A bain marie / mariae, or water bath.

Bezoardick – A medicine used as an antidote against poison and containing bezoar stone. Also said to be a promoter of sweat. The stones came from animal stomachs and could either be physical stones ingested by the animal, as well as indigestible hair or fibre balls trapped within the stomach.

Bolus – A mass or lump of prepared food for swallowing like a very large pill.

Carminative – Medicines allaying spasmodic bowel pain and dispelling flatulence.

Cataplasm – Essentially a poultice made with oils and fats into which herbal vegetable (or other) medicinal ingredients were mixed.

Clyster/Clister – An enema.

Compound Medicine – Medicines composed of several ingredients (see also Simples, below).

Cordial – Warm and stimulating medicines that raise the Spirits, and were also thought to strengthen the heart (derived from the Latin, *cordus*).

Corroborant – Possesses the power of strengthening, like a tonic.

Decoction – The liquid obtained by boiling vegetable substances in water for a time, to forcibly extract the soluble constituents. The solutions often contained solid matter that would settle out like a sediment, and so could be subject to fermentation in warm weather. Decoctions, therefore, had a limited shelf life though those made with wine lasted a little longer.

Deobstruent – Medicines having the power of removing any obstructions.

Diaphoretic – A medicine that causes increased perspiration, but gentler in its action than a sudorific (see below).

Electuary – A soft mass, consisting of a medicinal substance in powder form mixed with honey, or sugar syrup made with water or wine.

Epispastic – A substance or item applied to the skin which causes blistering and serous discharge. These days we would probably term this a vesicant.

Epitheme – A lotion, fomentation, or any external application.

Fomentation – Partial bathing with warm water, simple or medicated.

Humours/Humors – A concept, based on ancient medical writing, where the body was believed to have four *Humours*, blood, phlegm, black bile, and yellow bile, each having different qualities respectively: hot and moist, cold and moist, cold and dry, and hot and dry. Lifestyle and illness could put these Humours out of balance and it was the physician's job to re-balance them through his remedies and treatment of the patient.

Infusion – Differs from a decoction (above) in that the active principles of a substance (vegetable or mineral) are extracted by steeping in cold or hot water, rather than through active boiling. In the case of plants, vegetable material is bruised and allowed to stand immersed in water for a time. As with decoctions infusions have a short shelf life, varying from a few hours to a few days.

Inspissated Extracts – Although not mentioned in any of the original texts within this book, this is a term that will be encountered in old herbal recipes should you decide to research further. An inspissated juice is essentially the *sap* from vegetable matter that is bruised and then expressed by pressure, and afterwards filtered. Sometimes the juice was further evaporated to thicken it.

Menstruum – A solvent (oil, spirit, water, for example) used to extract the virtues of a herbal ingredient for use.

Miasma – A poisonous and contagious atmospheric vapour in which infective, disease causing, particles were suspended. Until the 'germ theory' about disease developed towards the end of the 19th century it was noxious miasmas which were thought to cause disease.

Oxymel – Honey and vinegar boiled to a syrup.

Phthisis – Pulmonary consumption, advanced tuberculosis / TB.

Plethora – A term generally applied to an over-abundance of blood (usually) in the system, the patient often appearing flushed. Physicians associated plethora with high living and lazy ways.

Simple/s – A medicine consisting of one substance, or only one active substance. The words were often used in relation to individual herbs having medicinal value when they were being discussed.

Spirits/Animal/Vital Spirits – This is quite a difficult one to describe succinctly in a short paragraph, but it was believed (based on Galenic medical theory) that the heart produced a life force or 'breath' (pneuma) that was conveyed to the brain via the blood and was subsequently converted into an animal spirit that travelled back through the nerves and muscles to give the human body motion and activity. It has an almost metaphysical dimension in some respects. So where you see reference to *Spirit* or *Spirits* in one of the quoted passages it is referring to this sort of 'life-force' concept and not the mindset of an individual, or alcohol.

Sudorific/k – An agent or medicine that induces sweating.

Theriac/a – This word frequently occurs in the herbal remedy vocabulary but is simply treacle or molasses. However, treacle was often at the heart of many compound remedies. *Theriaca Andromachi*, Venice Treacle, was composed of up to 70 ingredients and was used as an antidote against poisons.

Troches – Small little cakes of powders mixed with a gum (such as tragacanth) in solution, formed into round cakes, and then gently air-dried. The logic behind making troches was that the powdered ingredients would stay viable for longer since they were not exposed to spoiling air as they would be in powder form. Like pills troches were easily carried.

Vulnerary – A medicine or remedy which heals wounds.

ANATOMY OF A HERBAL RECIPE

Throughout this book there are examples of historically contemporaneous plague remedies, and at first glance some of them perhaps look unintelligible. What you are seeing in old herbal remedy books is, in effect, a medical *shorthand*; since those using the books on a daily basis would not need every word spelled out in full. However, once you understand the language behind the recipes then they are no longer a mystery and it becomes possible to dissect these remedies from centuries past. The other factor that becomes noticeable over the centuries is the rise of chemical pharmacy once chemical science begins to flourish from the 17ᵗʰ century onwards, and by the time of the Neligan-Macnamara *Medicines* [1864] British physicians are well and truly using remedies that extensively contain chemical ingredients.

To begin at the beginning of a remedy... When a recipe was written down or printed it was usual for the recipe to begin with an R or Rx which are simply shorthand for receipt, or recipe. Older handwritten Latin texts sometimes begin with the word 'Item'. All of these are traditional ways of identifying the start of a recipe and the instructions for making up the medicine.

Perhaps the next part of a recipe to understand are the measurements used. In terms of weight the following are typical (modern characters are used in the text of this book): Ə is a scruple, the character 3 which looks like the numeric 3 is a drachm or dram, and a sort of staggered three, ℥, is an ounce. The pound weight, *libra*, was represented by ℔ though on occasion it might be expressed as *lb*, *lib* or *li*. Other weight measurements are the *obulus* usually seen as *obol.*, and a *grain* expressed as *g*. Some ingredients were given by quantity, the two key ones being a handful, represented by an *m* or *M* (based on the Latin word for hand *manibus* or *manipulus*), and a *pugil* (from *pugno*, or fist) seen as a *p*. For liquid measures you may find *O*, denoting an *octarius* or eighth part of a gallon, in other words or a good old 'pint', and a minim is sometimes expressed with an ♏ or ♏ character. Drops of liquid are written as *gt* or *gut*, for a single drop, and *gtt.* for multiple drops, based on the latin word for a drop, *gutta*. An *f* in front of a quantity represented a fluid measurement, as in 'f℥', a fluid ounce.

Where a half part of a measure was required then it was followed by various characters that denoted *semis*, half. This might be expressed as s.s, a single 's', or when the long-S character is used f.s or fs, and in some cases by ß or ß.

By way of example: 3ij is two drachms
                                   3i one drachm
                                   3fs half a drachm
                                   Əij two scruples
                                   Əi one scruple
                                   Əfs a half scruple

Some other conventions used were:

Numero, N – a number of
Ana, ā or an. – of each
Partes æquales, part. aeq. or p.a – equal parts
Quantum sufficit, q.s. – sufficient quantity
Secundum artem, S.a. – the second art

The next part of any herbal recipe to translate are the actual ingredients used and, to a certain extent, if you understand a little Latin and know the scientific names of garden or wild plants then the task is much easier.

For example, *fol.* in a recipe is derived from the Latin word for a leaf, *folis*. *Sem.* comes from *semence* or the word for seed, while root or root material is seen as *rad.* from *radix*, the Latin word for root. *Succo*, is juice, and so on. Plant names too were frequently abbreviated, and generally based on their Latin names or common names that had been in use for generations.

Here are some real examples from an old book on *physick*:

Distilled water, thirteen ounces	Aq. destillat. Ʒxiij
Water, one pint and a half	Aquae purae, O iss.
Decoction of oak bark, one pint	Dec. quercus, O j.
Decoction of bark, twelve drachms	Dec. cinchon. 3 xij.
Bruised oak bark, one ounce	Quercus cort. contusi Ʒj.
Powdered rhubarb, one scruple and a half	Pulveris rhei, Э jss.
Tincture of squills, ten minims	Tinct. scillae, ♏x.
Fluid extract of dandelion, half a drachm	Ext. tarax. fluid, 3ss.
Extract of henbane, three grains	Ext. hyoscyami, gr. iij.

Put it all together and you will see something like the following anti-scorbutic recipe from *Pharmacopoeia Bateana* [1691]:

Serum Scorbuticum. Rx Fol. cichor. acetof. ā mxij. *abiet.* mvi. *cochlear. nafturt. aquat. ā* miv. *fem. coriandr.* Ʒi. *contufis adde fucc. aurant.* Ʒ4. *Seri lactis* ℔4. *facch. alb.* Ʒij. *clarific.*

What we have above then is a 'recipe' (the Rx) involving 12 handfuls each of chicory and sorrel leaves, 6 handfuls of fir or pine, of scurvy-grass and watercress 4 handfuls, and one ounce of coriander seeds. The instructions then indicate to bruise these ingredients then add orange juice, milk, white sugar, and then clarify.

Old printed medical books often used italics to emphasize the ingredients, or individual quantities or measurements in a remedy, so you may sometimes encounter an almost typographical yo-yo appearance in some of the remedies quoted in the pages that follow.

PLAGUE AS *WE* UNDERSTAND IT TODAY

For the purpose of this book the section that follows below provides the very broadest of overviews of *Yersinia pestis*, the bacterium at the heart of plague, since an outline knowledge of the nature of the plague pathogen will help you understand why successive generations of physicians failed to tame the plague, and the absolute futility of some of the remedies and measures employed in combating the disease. It should be said, however, that wherever a historic account of *'the plague'* or *'pestilence'* pops up over the centuries it cannot be assumed that *Y. pestis* was always the cause, although in the case of the Black Death and some other 14th century European plague outbreaks that does appear to be the case. So a little reading between the lines when that reference to the plague, pestilence, peste, or epidemia appears can be useful.

To somewhat muddy the waters of our contemporary understanding of the *Yersinia pestis* bacterium it should be pointed out that modern research has identified at least three different variants, biovars, of *Y. pestis* – Antiqua, Medievalis, and Orientalis – while *Y. pestis* is believed to have evolved from a parent species known as *Y. pseudotuberculosis*, and there are more than half a dozen *Y. pestis* strains. All of which has the scientific and historical communities interested in the plague locked in deep debate, but through the continued use of DNA testing of archaeological samples hopefully more of the jigsaw puzzle can be put together. Readers who would like to know more about the technical immunopathology and genetic aspects of the plague virus, its plasmid, antigen, receptor mutation and gene activity, can do no better than to look at the short article in JAPI by Clem, A. and Galwankar, S., entitled *Plague: A Decade Since the 1994 Outbreaks in India* [JAPI, Vol. 53, May 2005]. This contains an excellent short précis of technical aspects of the plague virus as a starting point plus a good list of references. For readers that really want to go into much more detail the World Health Organization produces a *Plague Manual* that contains a large amount of highly detailed information on the background, prevention and control of the disease, and which can be sourced digitally from the WHO website.

Although we tend to think of the plague as being a *disease of history*, it is very much present in remote areas of the tropics and subtropics among the rodent population. There the zoonoses causing the disease are transmitted by flea-bites between the rodent population, to other animals such as cats, mice and squirrels, and onwards to humans too. Fleas are the disease vector that successive genera-tions will be familiar with from their school history lessons on the Black Death pandemic, while rat populations are the reservoirs of *Yersinia* with some popu-lations being relatively resistant to infection themselves.

As recently as November-December 2014 plague broke out on the island of Madagascar, claiming about sixty lives, and in July 2014 the city of Yumen in the Chinese province of Gansu went into lock-down as a man died of bubonic plague after coming into contact with an infected dead marmot. Worldwide in 2013 there were 783 cases of plague, and 126 resultant deaths. Perhaps the severest plague scare in recent living memory was the 1994 epidemic in the Indian cities of Surat (Gujarat) and Beed (Maharashtra), although there is some dispute over the reported cases in Surat. However, panic took hold, and some sources say that more than a million people were displaced as they fled both the pneumonic and bubonic forms of plague (about 600 bubonic cases and around 150 pneumonic cases, resulting in some 50 or 60 deaths), while some countries imposed flight restrictions to and from India and others closed their borders to people and goods.

Credit for the discovery or identification of the pathogen causing plague belongs to two individuals working separately on the problem back in the 19th century; the Swiss-French physician Alexandre Yersin, and Japanese physician-bacteriologist Kitasato Shibasaburo, in 1894. Yersin released his scientific findings first and so the plague pathogen is named after him. Although Kitasato was working in exactly the same time-frame he appears to have been somewhat sluggish in informing the medical world about his discoveries, while some medical historians suggest that his scientific methodology was not totally sound. However, thanks to Kittasato and Yersin it was now possible for researchers to concentrate their efforts on finding ways of combating the *Yersinia* bacterium.

Among the initial findings were that *Y. pestis* was destroyed by high temperatures and desiccation (conversely we now know plague bacilli can remain viable for months, even years, if kept in darkness and prevented from totally drying out). While *Y. pestis* may be killed by heat modern references are at variance; according to one source 15 minutes exposure to 72°C will suffice, although other references suggest 55°C for 30 minutes, dry heat for at least 1 hour at 160-170°C, or moist heat at 121°C for at least 15 minutes. Sunlight, too, is destructive to the pathogen within a few hours, while ordinary chemical disinfectants containing chlorine were shown to be effective. Antibiotics were still an unknown and distant dream, but at least these scientific advances showed that there was potential to control plague, at least to a limited extent.

Inoculation against plague had first been hypothetically proposed in 1755, and again suggested in 1782 by Samoïlowitz, but it was only in 1897 that the first plague vaccine was developed. However it was later shown to be possibly ineffective in pneumonic plague, while causing adverse reactions. Indeed, adverse reactions have been an almost 'constant' in attempts to develop a vaccine, and even now there is no really safe and effective vaccine available against plague. Furthermore, the vaccine which is available does not protect individuals against the primary pneumonic form of plague, while it can take more than a month before a protective immune response develops. This makes the vaccine practically useless once a widescale plague epidemic or pandemic appears amongst a large population.

In relation to the modern antibiotic treatment of plague Streptomycin, Gentamicin, Tetracycline, Doxy- and Oxytetracycline can be effective, depending on the clinical form of the disease, while penicillins and cephalosporins are ineffective. Chloramphenicol is sometimes used in bubonic and septicaemic cases, and although sulfa drugs have been used they are not ideal and sometimes not by any means effective.

Make no mistake about it, *Yersinia pestis* is a potentially deadly and formidable opponent for medics, even modern-day ones with proper regulatory controls put in place. However, while a plague epidemic is not really an environment that you would wish to be in by choice, an understanding of the etiology, or causes, of the disease puts its dangers into a context that makes management and control of any outbreak possible, and largely explains the catastrophic, doom-laden, descriptions in old texts, and accounts as to why some people appeared to live normal lives amidst the medical mayhem that surrounded them.

Apart from the obvious infective, skin-puncturing, fleabite that we all know about from our schooldays, *Yersinia* can also enter the body through cuts, wounds and other skin lesions, and also through the soft mucous tissues of the mouth and nose, and even the conjunctival mucous membrane of the eyes. Once infected the disease may present itself in a number of clinical forms, *bubonic* and *pneumonic* being the most familiar ones. There are also *septicaemic* (septic), *pharangeal* (throat), and *meningeal* (attacking brain and spinal cord membrane) forms that sometimes develop as a secondary condition following on from primary infection symptoms.

In *bubonic* plague the infective *Yersinia* bacillus invades the body when an individual comes into contact with infected material (for example, bandages or clothes, or even a plague victim) where the infective material touches a skin lesion or mucous tissue. The onset of symptoms usually appears within a couple of days to a week and the victim develops painful swellings of the lymph nodes and lymph glands. Left untreated, mortality from the bubonic form can be as high as 60%, reducing to less than 15% where the antibiotics streptomycin and tetracycline are administered; streptomycin being regarded as best for the pneumonic form of the disease.

Complications occur where *Yersinia* bacilli then get into the victim's bloodstream producing the *septicaemic* form of the disease, which mostly results in toxic shock, and blood clotting of the small blood vessels that can cause secondary, often multiple, organ failure. The *septicaemic* form can also develop directly, from bacilli entering open wounds and getting into the bloodstream at the start.

The most deadly, contagious, clinical form of the disease is *pneumonic* plague that affects the lungs; where the victim develops severe pneumonia-like symptoms, a cough, chest pains, and breathing difficulty. The victim is infected through inhalation of the atomized respiratory droplets of another victim who has developed a cough. However, in itself *Yersinia* is not a true airborne disease, infecting the room air in which a patient resides, and by some accounts you would be relatively safe providing that you are two-plus metres away from a coughing victim. However, *pneumonic* plague will kill in less than 24 hours if antibiotics are not administered at disease onset, with symptoms appearing within one to three days of infection.

Bacilli lurking in infected sputum or body fluids that come into contact with skin lesions can cause the *bubonic* form of the disease but they may also cause development of the *pharyngeal* form that develops in the throat. Put simply, eating something contaminated with infected sputum or respiratory droplets will start the disease off, as can eating the undercooked meat of an infected animal – such as a squirrel caught in the wild.

The final, *meningeal*, form of the disease really develops as a secondary condition in the bubonic or septicaemic form of plague. It affects membranes covering the brain and spinal cord with the victim developing meningitis-like symptoms that may include delirium, confusion, and reduced consciousness to the point of coma. Indeed, many old accounts of plague victims mention symptoms such as these.

So, direct person-to-person transmission of plague does not generally occur (with the exception of the pneumonic form) without 'contact', and an understanding of the clinical forms that the disease takes should help you read between the lines of the descriptive texts in this book. In hindsight it is quite interesting to reflect that some of the physicians battling against the plague in pre-antibiotic times were quite close to the mark in their understanding and concept of the nature of plague. They simply lacked the analytical tools and technical know-how to deal with it and as a consequence millions of mortal souls died painful deaths in squalid conditions.

In 1665 John Gadbury questioned the rehearsed medical wisdom of his time; reflecting that there were healthy people surrounded by plague who remained unaffected by the outbreak. In his *London's Deliverance Predicted* he writes:

"*That which I here aim at, is to examine whether the* Peſtilence *be* infectious or catching *? If it be* infectious *and really* catching *in it ſelf; it muſt be ſo* equally to all perſons *that approach it, or that it approacheth; and this, either to* ſome degree of danger, *or elſe unto* death; *or elſe it muſt be* infectious *to ſame* particular perſons *onely* [sic].

If it be infectious *to all perſons, or catching to all alike; then* all perſons *that come into the* fight, *or within the* ſcent *of it, muſt neceſſarily be ſubject unto it; and this either unto* death, *or other leſſer degree of* danger. *There cannot be a* perſon, *either* man, woman, *or* child, *that is either ſhut up in a houſe with* perſons infected, *or that ſhall talk with any of them ſo ſhut up, though but at a* window, *or through a* Wicket, *but muſt be ſuppoſed to partake of the* infection; *for the* Talons *of a* Contagion *in this fence lay hold on them all. But how wide this is of the* Truth, *I leave to the judgements of any, that have their* five Sences *free from* infection, *and their* Reaſon *from* depravation.

In every great Peſt, *experience convinceth this opinion of* Error; *for in this great* City *we know (and ſee it daily now) that there are divers* perſons, *that have had (and yet have) the* Sickneſs, *the very* next door *unto them; on* both ſides *of them;* before *and* behind them; *and yet their* Perſons, Houſes, *and* Families *not ſo much as concerned in it, or touched with it. Many alſo are conſtantly viſiting their* Friends *and* Relations *that are viſited; yet (by Gods bleſſing) they remain* ſafe *and* found. And many *that I know (whoſe hard hap it hath been to be ſhut up (with others) in an* infected houſe, *out of which there have been ſeveral buried) yet, their good* fortune *hath been ſuch, that they have not only been freed from it, but have not had ſo much as a* head-ach *all that time; or in any conſiderable time afterward... How many are there of* Phyſicians, Chirurgeons, Apothecaries, Nurſes, &c. *that are daily among them, and yet eſcape not onely death, but the* diſeaſe *it ſelf?"*

In a book on European lazarettos the 18th century the English prison reformer John Howard [1791] also queried the notion of plague being airborne, and from personal experience at that: "*It may be aſked, how is it poſſible, if the plague be communicated by infected air, that a whole body of men in a town where it rages ſhould be capable of being preſerved from it, as is the caſe with Engliſhmen in Turkey; and alſo, why every individual in ſuch a town is not taken with it ? In anſwer to the firſt of theſe queſtions, it may be obſerved, that the infection in the air does not extend far from the infected object, but lurks chiefly, (like that near carrion) to the leeward of it. I am ſo aſſured of this, that I have not ſcrupled going, in the open air, to windward of a perſon ill of the plague and feeling his pulſe.*"

By the time of John Elliotson's *Principles and Practice of Medicine* in 1839 the contagious and infectious nature of plague was still being hotly debated: "*There can be no doubt whatever of its being a contagious disease but it is rarely communicated without contact. It is, for the most part, believed to be a* contagious *disease, in the strict sense of the word;- not* infectious. *One of the latest writers upon it [Mr Madden, a surgeon] says that if there be a deficiency of ventilation and cleanliness, so that the emanations from the patient are very much concentrated, it may be communicated by infection; but if there be any ventilation at all, then it can only be communicated by contact with the individual, or with something that he has touched. Some have denied, of course, that this disease is contagious but there are proofs without end that it is.*"

The work of the apothecary had its roots in archaic tradition; practitioners mixing and compounding ointments, potions and confections *according to art* for medicinal purposes. While more exotic ingredients might be imported from abroad local plant species were specifically grown in herb gardens, or harvested from the wild by the apothecary, their assistant, or professional pickers.

COMPREHENDING THE PLAGUE

From our position of medical and scientific knowledge in the early 21st century we can almost look upon the plague as just another disease, albeit one to be highly wary of. These days, when a new disease appears in our midst an army of scientists, researchers and academic institutions around the world will set to work putting the pathogen through a scientific wringer; the DNA or RNA dissected to find an Achilles' heel in the microbe and, hopefully, a cure or at least understanding of the nature of the bug as an outcome of all the research. The response of the medical community to the 2014-15 Ebola outbreak is a good example, though a positive conclusion to the research is still awaited.

In relation to plague, the key problem for populations and medics in the past was that they lived in a vacuum of understanding when it came to this dreadful disease. At the time of the Black Death, and for many centuries after, the populace was taught to fear God, and mostly certainly Satan, which kept the population in relative order for most of the time, and cowed dissent of thought and scientific exploration or questioning authority. This situation suited both Church and State who could profit from the subjected population while keeping them in a state of scientific darkness accompanied by threats of divine retribution should you step out of line or, woe betide you, sin.

At the time of the Black Death the devil, demons, and hell were part of people's belief systems, with many seeing the Pestilence and Plague as divine retribution for their misdeeds and sins.

Into this dark world place *Yersinia pestis* and the ravages it could cause; remembering that *bacteria* were not identified as a biological entity until the late 17th century by the Dutch scientist van Leeuwenhoek, and then the link between bacteria and disease figured out almost a century later, while *viruses* remained elusive until the end of the 19th-20th century divide.

Looking back in time medics certainly suspected something bothersome and small was behind the plague but it remained intangible. In the early 18th century Blackmore [1722] talks about the minuteness and hypothetical speed with which he believed the bugs travelled through the air: "*Thefe infectious Vapours or Exhalations diffus'd thro' the Air by their tenacious Conftitution adhere clofely to it, and therefore are carry'd to a great Diftance, and when with the infected Air they are by the Breath drawn into the Lungs, they affect the Spirits and the Mafs of Blood by their poifonous Quality... The Peftilential Atoms, which are the Seeds of the Plague, may likewife be communicated to the Body by Flesh-Meats, Herbs or Fruits on which the infected Air has fhed its Poifon, which being admitted into the Stomach, affect the Spirits, and create Putrefaction in the Blood."*

"*... the diftinguifhing Properties of the Plague are its Superior Contagion, and deftructive Quality, and that the laft confifts in the greateft Contrariety of Peftilential Vapours or Particles to the Animal Spirits, and the active Principles of the Blood, and that the firft is founded in greateft Minutenefs, Exaltation and Refinement, of thofe vapours by which they are able to pafs thro' the Air from Place to Place with extraordinary Velocity; and tho' thefe two Properties arife from the higheft putrefactive Power, yet the Idæas of one and the other are very different... yet the ready Conveyance of the Plague by the Air from Place to Place muft depend upon the extraordinary Smallnefs or Subtility of the Peftilential Matter; for no other Infectious Difeafe is communicated and convey'd in fo fwift a Manner, and to fuch a Diftance.*"

The mistaken belief that *Yersinia* is a truly airborne disease led George Pye [1721] – at exactly the same time – to suggest: "*... the Matter or Caufe of the Plague, whatever it be, muft neceffarily be received into, and conveyed by Air; and that therefore Walls and Lines cannot poffibly ftop or confine it.*" Pye was most certainly correct about walls, since they have no effect on the person-to-person trans- mission of the disease while cordon lines did, in fact, have some use in containing outbreaks of the disease, at least in some circumstances.

One of the most interesting contemporaneous accounts of attempts at quarantining with cordon lines can be found in the *History of Plague as it has Lately Appeared in the Islands of Malta &c.*, written by the British military physician J.D. Tully who was variously stationed in Malta and other Ionian islands during the early part of the 19th century. Published in 1821 the book is largely a journal of his attempts to isolate plague outbreaks on the islands. With some, but not total, success he and his team managed to keep many pockets of plague from developing into full-blown epidemics; manning cordon lines with troops to prevent movement of people. In a few cases the disease appeared behind the 'lines' without cause or warning and Tully admits they were at a loss to identify the provenance of the outbreak. However, in many cases the case source could be identified to an individual person and, much to his despair, this was usually down to people deliberately slipping through lines to visit relatives or loved ones in neighbouring villages, or indulge in a spot of smuggling around island coastlines, and so spreading the disease. In every instance though, Tully and his team would methodically track down contacts with almost forensic zeal, then isolate and monitor the linked individuals until they were either clear of plague or died. In one instance they had to follow up forty individuals who had been in contact with the primary case. Perhaps even more amazing is that very few of the team of soldiers under his command contracted plague despite their close proximity to numerous plague outbreaks. He does mention briefing his soldiers never to touch the individuals, or clothing and possessions of those who came near the lines, and describes one instance where he and the team – in the belief that they had been contaminated – strip naked, immediately burn their clothes, then scrub down with antiseptic solution. Given that this took place two hundred years ago it all has a very modern 'epidemic management' ring to it.

Despite the organized, logical, processes that medics such as Tully pursued they were still at a loss when it came to curing the plague. There is a sense of desperation lurking in the background of many medical texts written about plague over the centuries. Hancocke [1723], an 18th century Rector, comments: "*All Phyſicians confeſs, there is no Specifick yet found out, that will certainly take off and cure the Plague,*" while Sir Richard Blackmore [1722] is more expansive: "*There has no Antidote, or Specifick Medicine for the Cure of the Plague been yet found out by the moſt inquiſitive and ſagacious Phyſicians or Philoſophers; that is, there is not yet diſcover'd any Plant or Mineral, any ſingle of compounded Remedy apply'd outwardly, or taken inwardly, that is endow'd with a peculiar Quality ſo contrary to this Diſeaſe, that it will ſuddenly and certainly ſuppreſs and extinguiſh it. No Remedy is known in this Caſe like the* Jesuit's Powder *and* Opium, *which prevail by the Specifical Properties, one againſt an Ague and intermitted Fever, and the other againſt Pain and Wakefulneſs... Since then there are no Antidotes of certain Virtue for the Cure of the Peſtilence, the Perſons afflicted with that Diſtemper muſt be treated as thoſe who are ſeiz'd with Malignant Fevers...*"

Even as late as 1835, which we might like to think as being a slightly more enlightened age, Whitney pronounces in his American *Family Physician*: "*The plague is generally considered beyond the reach of medicine. Some authors, however, are firm in the belief that it does not differ materially from other summer fevers of a high grade, and that it may be prevented, and often cured.*"

Whitney's comment on treating plague fever as 'other summer fevers' was a not uncommon theme, not least because medics for many centuries tried to treat and cure a fever rather than its source – a pathogen. For example, flu may produce fever, so too typhoid and malaria, but they are all very different ailments so treating the symptom (fever) is not tackling the cause. Indeed, a large section of medical writing during the 17th and 18th centuries seems to have been devoted to trying to unravel the differences between 'types of fever', and a rather fatalistic view of plague fever is to be found in Hancocke [1723]: "*It is the moſt common Opinion of Phyſicians, that there is no Difference between the malignant Fevers and the Plague, but in Degree, in the Height of Infection, and Greater Contagion or Aptitude to infect others; and they make no other Diſtinction between them than this, that in other malignant Fevers more live than die, in the Plague more die than live.*" It should be pointed out that Hancocke was a Doctor of Divinity not a medical practitioner.

The comment below, from Blackmore again, shows how vague and woolly medical science was: "*... that Species of ill-condition'd Fevers, call'd Malignant, which is a Term, like many others in Phyſick, that has with many no determinate Meaning, nor any diſtinct Idea in the Mind anſwering to it. I ſhall therefore uſe the Word Malignant, as ſignifying ſome Degree of Putrefaction or entire Separation of ſome Parts of the Blood.*"

Bogged down with academic debates on fevers this aspect of medicine would remain locked in a rut for years. The debate on fever in relation to plague and other diseases is further exemplified by some of the following text extracts. For example, Thomas Willis [1685] exposes a key pitfall in relying upon fever as a pointer to identifying plague. Remembering that the rapid onset of plague can overwhelm the victim in a matter of hours Willis comments: "*... the Peſtilence by reaſon of its Symptoms, reſembling thoſe of a common Fever, we find our Dangers but too late...*" However, in the same treatise Willis separates out Malignant Fever, Plague fever, and also a Pestilential Fever, the latter being somewhat problematic since Pestilential Fever was regarded by many medics of the time as being pure and simply the Plague, with the Plague commonly referred to as The Pestilence. By way of explanation he says: "*Theſe Fevers differ from the Plague, and from each other, according to the Degrees and Vehemency of the Contagion and Deſtruction; ſo that the Plague is a Diſeaſe contagious, and deſtructive to Mankind in the higheſt Degree: A Peſtilential Fever is that which reigns with leſs Difuſſion of its Miaſm, and with leſs Mortality...*"

A few decades later Sir Richard Blackmore [1722] reflects that the highest forms of malignant fever are the same as those of lowest sort of plague fever, with it being hard to distinguish them, while informing his readers that: "*... Fevers do not differ in Nature and Eſſence but in Degree, and as in Morals ſuch a Superiority conſtitutes a new Species. Theſe different Species are the Putred, the Miliary, the Petechial, that is, the Fever accompanied with ſmall red or ſcarlet Spots, and that attended with blue, purple and blackiſh Ones diſpers'd in the Skin, and ſometimes, when more Malignant, with Mortifications in ſeveral Parts of the Body, and theſe ariſing from greater Putrefaction are often contagious and deſtructive.*" That reference to the '*Degree*' of a fever clearly underlines the medical thinking of the time, and why fever was a very blunt instrument in diagnosis of disease or an ailment. Incidentally, putrid fever was associated with what we would regard as a form of typhus, miliary fever is where sweat trapped in the epidermis causes irritation and red eruptions that resemble millet seed, while petechial fever is entirely related to the plague. Indeed, once a plague victim developed the unelevated purple-black *petechial* spots or lines on the skin that was regarded as a sure sign of imminent death, and so petechia were often called 'death tokens'. Often they merged to form larger darkened patches.

The cause of black and dark skin blotching was put down to clotting of the blood, the symptom of intravascular coagulation often occurring in plague attacks. Willis' *London Practice of Physic* [1685] provides us with some insight into 17th century thinking about the appearance of the blotches: "*... for the Spirits reſiding in both Liquors, eſpecially in the Blood, are no ſooner touched with the breath of a malignant Contagion, but a Coagulation is cauſed in the remaining Liquor, even as when Milk turns ſour, or has an acid Juice mixt with it: wherefore, Portions of it being greatly tainted with Venom, ſoon grow clotty, and like extravaſated Blood, fall into a Corruption with a Blackneſs; whence preſently they ſtop the Motion of the reſt of the Blood in the Heart and Veſſels, and coagulate it more by reaſon of their Ferment. Now whatſoever is gathered together into Clots by Coagulation, unleſs it be preſently caſt forth, brings Death in a ſhort time, by ſtopping the Circulation of the Blood; and being driven outward to the Circumference of the Body, is ſtopt in motion in the narrow Involutions of the Veſſels; and either being wholly deſtitute of Spirit, as tho it were Planet-ſtricken, it produces black or blew Marks by its Mortification, or by reaſon of the Salt and Sulphur exalted by the peſtilential Ferment...*"

The lack of clarity on clinical diagnosis of fevers, and the nuances between fever 'species', as it was sometimes termed, was certainly a definite drawback to rapid diagnosis of plague when and where it appeared, and that rapid diagnosis is important given the infectious nature of plague and the chance that the primary source may transmit the disease to others and thence to rapidly expand among a population. In our own time we have seen the difficulty of containing outbreaks of SARS, Bird Flu and Ebola, for example. But while inadequacies in clinical practice were one part of the problem of containing and successfully dealing with plague there were other equally important factors that coloured progress, particularly those of superstition and cultural beliefs. Again, this has been one of the main problems in containing the West African Ebola outbreak in 2014. There, parts of the population believe ebola is a 'curse' for evil misdoings, while the traditional funeral custom of physically embracing the dead has led to ebola contamination, and on top of that there is poor sanitation in many areas. Indeed, the spread of, and local reaction to Ebola in West Africa has many parallels with the European plague outbreaks that raged over the centuries.

Tully [1821], the British military surgeon, reported that in Corfu where he had been stationed and the plague had struck, the local population '*over whom ignorance and superstition seemed to reign with unbounded sway*' attributed:

"... the whole of the evils, with which they were afflicted, to the agency of a spirit, being that of a man who had been murdered in the neighbourhood of the village some months before. They were confident that this was the true cause of their sufferings, and endeavoured to make every atonement to the angry spirit by means of church offerings, prayers, and processions. All who died, it was asserted, had been attacked either in the evening or returning from their field labours or during the night. They believed that the spirit inflicted punishment by stripes, and by efforts at strangulation, and that the terror excited in the minds of all those who were attacked, and the continued nightly persecution of the spirit hurried them from one extreme of agony to another until their sufferings terminated in death." The reference above to 'punishment by stripes' is probably to vibices, linear purple streaks on the skin caused by subcutaneous effusion of blood and often a symptom of a plague attack along with petechiae.

Shortly after the Black Death had done its worst in Europe the College of Physicians in Paris investigated the causes of the Black Death for the King and pronounced in a consilium, in late 1348, their opinions on ways of avoiding or preventing plague. The final recommendations would have achieved precious little but because this official 'Opinion' came from so learned a body it remained largely uncontested, and continued like a bibliographic echo to pervade the public mind-set as well as medical thinking for many centuries, even among some more enlightened medics.

Now that we know, for example, that desiccation and sunlight can kill Yersinia there is some small sense in their notion of avoiding dark, damp, rainy, environments where the the bacterium would still be able to exist. However many of the observations and recommendations are pretty far-fetched and, despite the Faculty's insistence that they put hard fact before astrology (which still lurks in the background), parts of the 'Opinion' are based on texts from hundreds of years before:

"We, the Members of the College of Physicians, of Paris, have, after mature consideration and consultation on the present mortality, collected the advice of our old masters in the art, and intend to make known the causes of this pestilence, more clearly than could be done according to the rules and principles of astrology and natural science; we, therefore, declare as follows :—

It is known that in India, and the vicinity of the Great Sea, the constellations which combated the rays of the sun, and the warmth of the heavenly fire, exerted their power especially against that sea, and struggled violently with its waters. Hence, vapours often originate which envelope the sun, and convert his light into darkness. These vapours alternately rose and fell for twenty-eight days; but at last, sun and fire acted so powerfully upon the sea, that they attracted a great portion of it to themselves, and the waters of the ocean arose in the form of vapour; thereby the waters were in some parts, so corrupted, that the fish which they contained, died. These corrupted waters, however, the heat of the sun could not consume, neither could other wholesome water, hail or snow, and dew, originate therefrom. On the contrary, this vapour spread itself through the air in many places on the earth, and enveloped them in fog."

"Such was the case all over Arabia, in a part of India; in Crete; in the plains and valleys of Macedonia; in Hungary; Albania and Sicily. Should the same thing occur in Sardinia, not a man will be left alive; and the like will continue, so long as the sun remains in the sign of Leo, on all the islands and adjoining countries to which this corrupted sea-wind extends, or has already extended from India. If the inhabitants of those parts do not employ and adhere to the following, or similar means and precepts, we announce to them inevitable death — except the grace of Christ preserve their lives."

"We are of opinion, that the constellations, with the aid of Nature, strive, by virtue of their divine might, to protect and heal the human race; and to this end, in union with the rays of the sun, acting-through the power of fire, endeavour to break through the mist. Accordingly, within the next ten days, and until the 17th of the ensuing month of July, this mist will be converted into a stinking deleterious rain, whereby the air will be much purified. Now, as soon as this rain announces itself, by thunder or hail, every one of you should protect himself from the air; and, as well before as after the rain, kindle a large fire of vine-wood, green laurel, or other green wood; wormwood and chamomile should also be burnt in great quantity in the market places, in other densely inhabited localities, and in the houses. Until the earth is again completely dry, and for three days afterwards, no one ought to go abroad in the fields. During this time the diet should be simple, and people should be cautious in avoiding exposure in the cool of the evening, at night, and in the morning. Poultry and water-fowl, young pork, old beef, and fat meat, in general, should not be eaten; but on the contrary, meat of a proper age, of a warm and dry nature, by no means, however, heating and exciting. Broth should be taken, seasoned with ground pepper, ginger and cloves, especially by those who are accustomed to live temperately, and are yet choice in their diet. Sleep in the day-time is detrimental; it should be taken at night until sun-rise, or somewhat longer. At breakfast, one should drink little; supper should be taken an hour before sun-set, when more may be drunk than in the morning. Clear light wine, mixed with a fifth or sixth part of water, should be used as a beverage. Dried or fresh fruits with wine are not injurious; but highly so without it. Beet-root and other vegetables, whether eaten pickled or fresh, are hurtful; on the contrary, spicy pot-herbs, as sage or rosemary, are wholesome. Cold, moist, watery food is, in general, prejudicial. Going out at night, and even until three o'clock in the morning, is dangerous, on account of the dew. Only small river fish should be used. Too much exercise is hurtful. The body should be kept warmer than usual, and thus protected from moisture and cold. Rain-water must not be employed in cooking, and every one should guard against exposure to wet weather. If it rain, a little fine treacle should be taken after dinner. Fat people should not sit in the sunshine. Good clear wine should be selected and drunk often, but in small quantities, by day. Olive oil, as an article of food, is fatal. Equally injurious are fasting or excessive abstemiousness, anxiety of mind, anger, and excessive drinking. Young people, in autumn especially, must abstain from all these things, if they do not wish to run a risk of dying of dysentery. In order to keep the body properly open, an enema, or some other simple means, should be employed, when necessary. Bathing is injurious. Men must preserve chastity as they value their lives. Every one should impress this on his recollection, but especially those who reside on the coast, or upon an island into which the noxious wind has penetrated."

While the Paris faculty may have denied falling back on astrology and similar disciplines others were of entirely the opposite opinion, as shown in the following 15th century manuscript text. It comes from an English translation of a treatise on the pestilence (sometimes known as *De Pestilentia Liber*) by John of Burgundy, a physician in Liege, and is believed to have been written in the 1360s. There are numerous Latin, French and English versions of the original, some longer than others, from around the period. The extract here is from Sloane MS 3449 held in the British Library, but what is important from the medical point of view is the suggestion that readers should not entrust their health to any doctor lacking astronomical knowledge:

"*Also alle they whos complexion contrary to the aire that is chaunged or corupte abiden hole and elles alle folke fhuld corupte and dye at onys. The aire therfore fo corupt and chaunged bredith and engendreth in diverfe folks difeafe fikenes and fores.*
After the variauncez or diverfitees of theire humors for any worcher or evry thing that werechith performeth his werke after the abilite and difpoficon of the matier that he werkith ynne.
And by caufe that ther have ben many grete maiftirs and ferre lernyd in theoric or fpeculacion and groundly in fight of medecyne but they bene but litill proued in practik and therto alle-fully ignorant in the fience of Aftronomy the whiche fcience is in phifik wonder nedefull as witteneffith ypocras in Epidimia fua feying what phifician that ever he be and he kan not [know] aftronomy [,] no wyfe man owt to putte hym in his handis [,] for why aftronomye and phifik rectifien yche othr in effect and alfo that one fcience fheweth forthe many thynges hidde in the other for alle thynges in one thyng may not be declared.
And I 40 yere and more have oftyn tymes proved in practife that a medecyn gyven contrary to the conftellacion all thogh hit were both wele compownyd or medled and ordynatly wroght after the fcience of phifik yit it wroght nowther aftur the purpofe of the worcher nor to the profite of the pacient. And when fome men have gyven a medecyn laxatyf to purge downeward the pacient hath caften it out ayene above all thogh he lothed it noght.
Wherfore they that have not dronkyn of that fwete drynke of Aftronomye mowe putte to thife peftilentiall fores no perfite remedie for bicaufe that they knowe not the caufe and the qualite of the fikeneffe they may not hele it as feith the prynce of phifik Avicenna.

How fchuldest thou he faith hele a fore and thou knowe not the caufe. iij canone capitulo de curis febrium. *He that knowith nat the caufe hit is onpoffible that he hele the fikenes. The comentour alfo fuper fecundum phificorum feith thus[.] A man knowith nat a thyng but if he knowe the caufe both ferre and nygh. Sithen therfor the hevenly or firmamentall bodies bene of the firft and primytif caufes it is bihovefull to have the knowlechyng of hem for yf the firft and primytif caufes be vnknowen we may not come to know the caufes secondary. Sithen therfor the first caufe bryngeth in more plentevoufly his effecte than doth the caufe fecondary as hit fhewith.* primo de caufis. *Therfor it fhewith wele that without Aftronomye litill vayleth phifik for many man is periffhed in defawte of his councelour."*

The belief that a physician had to be acquainted with the heavens is reflected in Bullein's *Dialogue* [1578]. This populist book, if it can be categorized as such, is laid out much like a conversation between two people, one of them a physician. Upon being asked what knowledge of the *natural world* a physician should have, he comments:

"Most chiefly, for where as the Philosopher doth leaue, there the Phisition doth begin; that is, he must be first a good natural Philosopher, he must haue the knowledge of tymes and seasons, and bee acquainted with complexions of men, obseruyng the nature of thynges, and the climates vnder heauen, with the course of the Sunne, Moone, and Starres, ayre and diet, &c.," while on the the 'signs' of forthcoming pestilence he advises:

"The signes are moste manifest, whiche are the starres running course or rase after their causes. Oh, the most fearefull Eclipses of the Sunne and Moone, those heaueuly bodies, are manifest signes of the pestilence emong men, and the starres cadente in the beginnyng of Haruest or in the moneth of September; or muche Southe Winde or Easte winde in the Canicular daies, with stormes and cloudes, and verie colde nightes and extreame hotte daies, and muche chaunge of weather in a little time; or when birdes do forsake their egges, flies or thinges bredyng vnder the ground do flie high by swarmes into the ayre, or death of fishe or cattell, or any dearth goyng before, these are the signes of the Pestilence and euident presages of the same."

Southerly winds appear to have been a major preoccupation on the weather front with Johannes Jacobi in *Regimen Contra Pestilentiam* [1483] saying: *"... whanne grete wyndes paffen out of the fouth they be foule & unclene therfore whan thefe tokens appere it is to drede grete Peftilence but god in his mercy wil remeue it."*

In 1569 John Vandernote typically reflects this in the following century too: *"... and the fouthe winde whiche naturally is* infective. *And therefore fhall every perfon kepe the windowes locked in the morninge till eyght of the clocke ftanding againft the fouth."*

Contamination of the air is a common theme following the 14th century Paris faculty text and Vandernote says that among the 'tokens' (or signs) portending plague: "*The fourth* [ED – ie. sign], *when ỹ ſtarres are like the ofte to fall or flippe, this is a token that the ayre is inflamed, and full of venimous vapours.*" By falling and slipping stars I am wondering if that is a comment regarding comets and shooting stars.

Around fifty years later Thomas Thayre in his *Excellent Treatise of the Plague* [1625] still adheres to the celestial-air connection: "*I finde many cauſes that may corrupt the aire, all of which I will compoſe or include in theſe two....*
The firſt cauſe whereby the aire may be corrupted, is through the vnholeſome influence of the planets; who by their malitious diſpoſition, qualitie, and operations, diſtemper alter and corrupt the aire, making it vnwholeſome vnto humane nature...," and a little later on: "*I omit to write what I haue read concerning the alterations and mutations, that are ſometimes cauſed by the ſuperior bodies or planets here below vpon the earth; for vnto the learned it were ſuperfluous, & vnto the vulgar or common ſort, it would rather breede admiration then* [sic] *credit; but this euery man is vnderſtand,* Deus regit Aſtra, *God rules the ſtarres, and yet I doubt not, but through the Eclipſes, Exaltation, Coniunctions, and aſpects of the Planets, the aire may be corrupted, and made vnwholſome ſometimes, in ſomuch that diuers griefes are bred thereby.*"

Another short contemporary book from the same period, and which could perhaps be termed 'populist', is Stephen Bradwell's *Physick for the Sicknesse, Commonly called the Plague* in which he collates sources of information '*from choicest authors and confirmed with good experience*'. Quite whether Bradwell was an apothecary I am uncertain but there is a short catalogue at the end of the book extolling the virtues of potions that he sells, while he peddles the following advice to his readers concerning the 'The Putrid Plage' [sic]:

"*This comes of Putrefaction of the Bloud and Humors in the Body... This Putrifaction may be cauſed by the Influence of the* Starres, *who doe undoubtedly worke upon all ſublunarie bodies. For* Aſtrologers *are of opinion, that if* Saturne *and* Mars *have dominion (eſpecially under* Aries, Sagitarius, *and* Capricorn,*) a* Peſtilence *is ſhortly to be expected.*
... Now if the Starres *be peſtilently bent againſt us, neyther* Arts *nor* Armes, Perfumes *nor* Prayers, *can prevaile with them, who have neyther pitie nor ſenſe, nor power to alter their appointed motions.*"

While celestial happenings were regarded as partly responsible for the appearance of plague so too was the weather and climate, since these could alter the quality of air. With our understanding of meteorology today, particularly through weather and climate monitoring, we rather take it as read that changes in atmospheric conditions can change the quality of the air. There is nothing divine or mysterious about the appearance of a pea soup fog, smog, or cold damp mists hanging around.

In the past such conditions had an ominous dimension linked to them which is reflected throughout medical texts on plague. As most of us know all too well cold damp air presents an ideal springboard for a winter chill so perhaps these were also good conditions for *Yersinia* to thrive. But *Yersinia* is not a truly airborne bacteria and so would not generally have been lurking in wind blown clouds or mists, though certainly the cold damp conditions are quite the reverse of two of *Yersinia's* key weaknesses, dessication and sunlight.

In his populist *Dialogue* [1578] William Bullein reverts back to Hippocrates upon being asked what is the cause of the Pestilence: "*The Pestilent feuer, saieth Hypocrates, is in twoo* [sic] *partes considered; the first is common to euery man by the corruption of* [the] *ayre; The second is priuate or particular to some men through euil diete, repletion, whiche bringeth putrifaction, and finally mortification...*"

A little further on Bullein connects air quality with water, but the notion of bad air contributing to plague is evident: "*When there doth come a sodaine alteration or change in the qualitie of water from Colde to heate, or transmutation from sweetenesse to stincke, as it chaunceth waters through corrupted mixture of putrified vapour*[s] *infectyng bothe ayre and water, whiche of their owne simplicitie are cleane, but through euill mixture are poysoned; or when stronge Windes doe carrie pestilent fume or vapours from stinkyng places to the cleane partes, as bodies dead of the vnburied, Or mortalitie in battaile, death of cattell, rotten Fennes, commyng sodainly by the impression of aire, creepyng to the harte, corruptyng the spirites, this is a dispersed Pestilence by the inspiration of* [the] *ayre.*"

In his *Defensative against the Plague* [1593] Kellwaye reaches even further back in time, to Avicenna almost five hundred years previous. For advice on *Warninges of the plague to come* Kellwaye tells his readers: "*Avicen a noble Phyſtion ſaith, that when wee ſee the naturall courſe of the ayre, and ſeaſons of the yeere to be altered, aſ when the ſpringe time is colde clowdie, and drie, the harueſt time ſtormie and tempeſtuous, the morninges and evenings, to be very colde, and at noone extreame hote, theſe doe foreſhew the plague to come.*"

Air contaminated by 'matter' arising from the stagnating environments or the earth is a common thread through many plague texts down the centuries and Thayre [1625] alludes to this when he writes: "*The ſecond cauſe, whereby the aire may be corrupted, is a venemous euapuration ariſing from the earth: and from fennes, moores, ſtanding muddy waters, and ſtinking ditches and priuies, or ſome dead bodies vnburied, ſtinking channels... and multitudes of people liuing in ſmall and little roome, and vncleanely kept; all theſe are cauſes and meanes whereby the aire may be corrupted.*"

The author of *Charitable Pestmaster* [1641] was keen to keep his readers away from foul smelling effluvia lest they contaminate the body: "*Therefore whoſoever would preſerve their bodies from infection, let them firſt make their peace with God, in whoſe hand is the power of life and death. Then let them uſe the meanes, and ſhun all those things that are able to beget this diſeaſe, as all infected and corrupted aire, all fogges and miſts that do ariſe from the earth or water, and all ſtinking ſmels that do ariſe from dunghills, ſinks, graves, carrion, ſnuffs of candles, or rotten fruits, or any thing elſe that doth putrifie and ſtink.*"

When we get to the 18th century air is still regarded as a key link in the dissemination of plague. In a *Practical Treatise of the Plague* [1720] Joseph Browne addresses Dr. Mead in his opening introduction (Mead being an important English authority on the plague at the time), suggesting that there were things other than corrupted air responsible for plague. What the extract from that book does show is how air was regarded as the *primary factor* behind the plague: "*You very rightly obſerve in the Beginning of your Diſcourſe, that Contagion is propagated by three Cauſes; the Air, diſeaſed Perſons, and Goods tranſported from infected Places; and with Submiſſion to the Judgment of the Learned, if you had added Two other Cauſes, viz. Diet, and Diſeaſes that are the Cauſes of other Diſeaſes; there are many Authorities, and a great deal of Reaſon to back them. For though the Air is univerſally allowed to be the firſt Cauſe in propagating Contagion, becauſe being too moiſt when Showers deſcend during the ſultry Heats of* Auguſt, *eſpecially in hot Countries near the Sea in* Africa, *and* Grand Cairo *in* Egypt, *and in other low Scituations [sic], thereby the Spring of the Fibres is more abundantly relaxed, which makes the Circulation flow, and from that Slowneſs renders the Perſpirations languid, whence the Humours become apt for Corruption, one of the grand Cauſes of Contagion: Beſides, the Regulation or Obſervance of the external Air, is neceſſary to the Body, as it hinders too great an Expanſion of the Fluids and Solids; from whence it may be eaſily judged, that a depraved Air is the Author of malignant Fevers.*"

Despite the insistence of some physicians that contaminated air was significantly involved in spreading plague there were accounts that should have cast doubt. In an English translation of the account by Charles de Mertens [1799] of the 1771 Moscow plague outbreak we are told: "*When we viſited any of the ſick we went ſo near to them that frequently there was not more than a foot's diſtance between them and us; and although we uſed no other precaution but that of not touching their bodies, clothes, or beds, we eſcaped infection.*"

The notion that air was responsible for the plague problem carried through into the next century too. In the American work *The Treasure of Health* [1819] which can only be regarded as a populist 'home help' and quack book rather than a learned tome, Lewis Merlin sources his content from French and other languages. Indeed, when you consider that among his remedies for plague were two using dried toads, then you get the measure of his book, however it does reflect the knowledge-base or understanding that the American general public would have had at the time. In the section *'Of the Plague; and other Contagious Diseases, and many Sovereign Remedies, always used with the greatest success,'* this author tells his readers: *"The plague is nothing less than a very contagious and epidemical disease, which springs from a venomous exhalation contained in the air, and after-wards encreased* [sic] *through contagion, which attack the human race insidiously, and exposes their lives to danger."*

Contrary to the American work above there were at least some who had seen the light. In 1813 the English physician Richard Pearson wrote a book on the plague for the use of his fellow practitioners across Britain, particularly those working in sea-ports. These practitioners, he believed, were more likely to see appearances of the disease through oceanic travel and trade, and tells his readers categorically that: *"The contagion of the plague is not communicated by the atmosphere, but, as before stated, by contact. It should not, however, be understood, that it is communicated by contact alone, and in no other way. When the tongue, fauces, stomach, or lungs, are affected with carbuncle, the breath, at the moment it is emitted, and in the immediate proximity of the mouth, may be capable of producing infection; but in other cases, and which are by far the most numerous, where those internal parts are not diseased, it is highly probable that the contagion of the plague is not propagated by the breath."*

While Pearson is obviously admitting that there is potential for airborne transmission in the immediate vicinity of a plague victim – possibly leading to the pneumonic plague form in someone standing nearby – he is ruling out the concept of general airborne transmission of the disease. Around this period, from the late 18th century to the end of the 19th century, a lot of data was collected by English and French military doctors involved in various foreign campaigns. They came upon plague on a not infrequent basis, and made attempts in a logical and rational manner to conquer the disease, albeit unsuccessful.

Another frequent 'link' to perceived bad air quality, and its connection with the appearance of plague, was that of fumes from volcanoes and other similar sub-terranean eruptions, which seems somewhat inconceivable these days. In the posthumously published *London Practice of Physic* [1685] based on Thomas Willis' writings we hear that: *"Moreover Earth-quakes, and freſh-opened Grotto's and Caverns upon the cleaving of the Earth, by reaſon of the Eruptions of a malignant and venemous Air, have often given Beginnings to Plagues,"* but it is unclear whether those thoughts belong to the book's Editor or to Willis. To a certain extent that latter point is immaterial since the inclusion of the general idea shows that the concept was contemporaneous to the period.

And although he acknowledges the natural toxicty of volcanic sulphur vents in the following passage, Blackmore [1722] is quite expansive, perhaps even flowery, about the volcanic connection with plague; but again it reflects a widely held belief among plague observers of the past: "... *Plagues are often bred in the Bowels of the Earth, while the Reeks and Fumes of various Kinds arifing from the Strife and Conflicts of fermenting Minerals and unripe Metals, and agitated by Fires that rage in the Vaults underground, having fill'd their Caverns, and being fet on Fire by their own ftruggle, or force neighbouring Flames, and wanting Room, like kindled Gunpowder, to diffufe it felf, burft their Prifons by furious Earthquakes, and break thro' the Chafms and Difruptions of the Ground in violent and contagious Tempefts: And thefe fill the Regions of the Air with crude, peftilential Seeds and Subterranean Poifon, which malignant Eructations gathering to themfelves the hurtful Particles which they meet with in their Way, gain greater Force, and being drawn into the Lungs by the Breath, infect the Vitals, and execute their terrible Tragedy."*

The anonymous author of the short booklet *Some Observations Concerning the Plague* [1721] mentions that Robert Boyle, the celebrated chemist and man of science no less, had his own thoughts on the subject: "*As to Mr. Boyle's Opinion concerning the Rife of the Ordinary Plague it is in fhort* This: *He inclines to think that the* Malignant Difpofition *of the* Air *whereby the* Plague *is propagated, if not firft produced, is imputable to fome Kind of* Subterraneal Expirations, *and particularly to* Arfenical *Fumes; but as to this He is far from being Pofitive, or Dogmatical...*" That final comment about Boyle being unsure or dogmatic is perhaps what you would expect from a man of science who puts proof as a cornerstone of scientific inquiry.

The notion that plague could possibly come from the bowels of the planet gave rise to another phenomenon in the historical medical debate on the sporadic and spontaneous appearance of the plague – that of the '*localists* vs *contagionists*'. Quite simply the contagionists believed that new plague outbreaks were brought to Britain from without; whether it be from Europe or further afield, and generally appearing in English ports first. The localists argued that the source of plague must remain lurking in the soil, be it in the insanitary conditions of large urban areas, sewers, or unhealthy swamps. In other words, an endemic pool of the plague pathogen just waiting to be re-ignited from time to time when conditions were right. Understanding what we do today about plague we can see that there was a grain of truth in both the localist and contagionist camps. That 'pool' would be any locally infected rodent population – in whatever affected part of the world – which can harbour the plague pathogen long-term and then transfer it to humans through fleabites or physical contact with plague-carrying animals. Meanwhile the *contagionists* were obviously right about the plague travelling by means of human carriers, though only partially correct on the contaminated goods front (*Yersinia pestis* being susceptible to heat, sunlight, and dessication, although with enough moisture present it remains viable for a good while).

For devout religious citizens a personal contributor to becoming infected with plague, and placed ahead of corrupted air as a key cause, was sin. This pops up in both the 16th and 17th centuries, periods when religion was an important part of the politics of daily life. In *Governance and Preservation of them that Feare the Plage* [1569] the physician-surgeon John Vandernote admonishes his readers: "... *let them abſtain of all thynges cauſing putrefaction, aſmuch* [sic] *as is poſſible, like as is ouermuch copulacion between man & woman...*" and a little later, "... *lyke as in them that are greate medlers in Lechery.*"

In another section specifically warning '*Of carnall copulacion*' he goes even further, suggesting marriage should be put off: "*You muſt abſtaine from to* [sic] *much carnall copulacion, and therefore in that ſeaſon no man ſhall mary, neither come or medle with many women, for thorow the ſame ſhall you be inclined unto the corruption wherof the Plage doeth come, you ſhall live honeſtly, and ſpecially thoſe that be corrupte or inclined to the ſame ſickneſſe.*"

The French, too, expressed warnings along similar lines, Jean Vigier in his *Traicte de Peste* [1614] advising readers: "*La compagnie des femmes ou acte venerien eſt merueilleuſement dangereux en temps de peſte.*"

In England sin and corrupt air were equal dangers, Thomas Thayre [1625], saying that among the causes of plague: "*The firſt and chiefeſt is ſinne. The ſecond is the corruption of the aire,*" while Thomas Sherwood [1641] follows the same tune: "*There are divers cauſes of this diſeaſe. The firſt is ſin, which ought to be repented of. The ſecond is infected and corrupted air, which ſhould be avoided.*"

The hordes of unwashed peasantry also came into the blame game for the arrival of plague; hardly fair when one considers there were not many options for those who had only an earth floor to sleep on, perhaps sprinkled with rushes or a little straw or leaves if they were fortunate.

Responding to a question as to who is visited by pestilence the physician in Bullein's Dialogue [1564] replies: "*Moſte chiefly to them vnder the place infected, then to sluttiſhe, beaſtly people, that keepe their houses and lodynges vncleane, their meate, drinke, and clothyng moſte noyſome, their laboure and trauaile immoderate; or to theim whiche lacke prouident wiſedome to preuente the ſame by good diete, ayre, medicine, &c.; or to the bodies hotte and moyſte; and these bodies do infect other cleane bodies, and whereas many people doe dwell on heapes together...*"

The same sentiment is found in Bradwell's *Physick for the Sickneſſe, Commonly called the Plague* [1636] published in the same year that there was a serious plague outbreak in London: "Poore people *(by reaſon of their great want) living ſluttiſhly, and feeding naſtily and unwholſomly, on any food they can with leaſt coſt purchaſe, have corrupted bodyes, and of all others are therefore more ſubject to this* Sickneſſe."

Both of the two texts above also lead on to another main suspect in the causes of plague, that of diet. In fact there are quite large sections on diet in old plague treatises that catalogue what should and should not be eaten during the time of plague, and other foods that were believed to engender or make a personal visitation by the plague more likely. On the other hand many of the poor had little food and so would have been naturally weaker, and therefore potentially more susceptible to catching diseases and in ill health generally.

So, in his populist *Dialogue* [1564] William Bullein includes poor food among causes of plague: "*Also by repletion, Venus, Bathyng, or opening the poures, rotten foode, fruite, much wine, or immoderate labour, or the tyme beyng hotte and moyſte. These are greate cauſes.*"

After sin and corrupted air as causes Thayre [1625] comments: "*The third and laſt cauſe, is the euill diſpoſition of the bodie, bred by euill diet...*" And again you see the sentiment in Sherwood [1641]: "*The third and evill* [sic] *diet, which ſhould be amended. The fourth are evill humours heaped together in the body, being apt to putrifie, and beget a Fever, which muſt be taken away be convenient medicines.*"

In the re-package of Thomas Willis' works, *London Practice of Physic* [1685], among the 'signs' of a potential plague visitation there are some familiar themes present such as damp air, but there is also the mention of famine and dearth of food. These last two items have their roots in texts that precede the 14th century Paris *concilium*, this 17th century text telling us: "*There are a great many Signs occurring to us, which fore-ſhew that the Peſtilence will happen in a ſhort time, to wit, if the Year does not keep its Temperament, but has immoderate and very unſeaſonable exceſſes of Heat or Cold, Drought or Moiſture: if the Meaſles or Small Pox are every where very rife, if Phlegmons, or Bubo's accompany reigning Fevers; from a proceeding Famine a moſt certain Preſage is taken of an enſuing Plague; for the like Conſtitution of the Year which for the moſt part brings a Dearth of Proviſions, by reaſon of the Corn being blighted, is apt alſo to produce a Plague; alſo the evil ſort of Dyet, which ſuch as are preſt with Hunger make uſe of, eating all kinds of unwholſome things without choice, diſpoſes their Bodies more readily to receive the Contagion.*"

By the time of Browne's *Practical Treatise of the Plague* [1720] medics are convinced there are connections between food and disease: "*Now tho' the Air, as I ſaid before, be univerſally allow'd a Cauſe; the Aliment oꝛDiet, becauſe it affords Matter to the Juices, does not leſs contribute to the Generation of Diſeaſes; for the more a Man eats, the leſs he perſpires; the leſs he perſpires, the more Danger there is of a* Plethora. *Again, all Things that ſubject to Fermentation are bad, and all Things* which relax the Tone, and incline it to Flatulencies, Diarrhea's [sic], and all putrid Diſeaſes which ariſe from too great Plenty of Serum."

Returning to an earlier period in time John Jones [1566] has this: "The dyet that ſooneſt cauſeth corruptiō, is that which is receyued by meats of yl noriſhment, as fruites, wynes, herbes, and al others of that kindes, alſo famine is euyl to be ſuſteined. For the powers being to [sic] open, is made the readier to receiue infection: and alſo after ſedinge to groſſly corrupteth the body." Particularly interesting there is that phrase 'the powers being to open', or skin pores being opened, allowing ingress of contagion.

Again, the idea that bad diet could store up, or produce, a state of being that made infection from plague more likely is fully expanded upon by Thayre [1625] which gives an understanding into the thought process behind good and bad diet and its' effect on the Humours: "The third cauſe of the peſtilence, is the euill diſpoſition of the body, which is bred by euill dyet, the body being repleat with corrupt and ſuperfluous humors, which humors be ready to putrifie and rot vpon any light occaſion, and when ſuch a perſon doth but receiue into his body by inſpiration, the corrupted and infectious aire, he is therewith by and by infected, his body being diſpoſed thereunto through ſuperfluous and corrupt humors abounding, whereas contrary wiſe a body of a good diſpoſition, I meane a body free from groſſe, corrupt, and ſuperfluous humours, is not eaſily or lightly infected becauſe there is not that matter for the infectious ayre to worke vpon. And againe, nature is more ſtrong to repell the infectious or corrupted ayre, if it be receiued, and this is the cauſe why one perſon is rather infected then another, namely, the diſpoſition of the body." So, bad food and poor diet created bad 'humours' that could be infected, and the best way to deal with that was to eat recommended foods, both during plague times but also at other times.

In 1593 Simon Kellwaye in his Defensative refers back to a diet suggested by Hippocrates who lived around the fourth century BC, so pre-dating Kellwaye's work by very nearly two thousand years. According to the Hippocrates-Kellwaye diet – and it should be said, championed by many others over the centuries too – only easily digestible meats were permissible, so: "Cockes, Capons, Hennes, Pullets, Partridge, Feaſants, Quayles, Pigeons, Rabbets, Kydde, Veale, Mutton, Birdes of the mountaines, and such like," were acceptable, while "Beefe, Porcke, Venison, Hare, and Goates flesh" were to be avoided, and "water foules... Ducke, Swanne, Gooſe, Widgen, Teale and such like..." also off the list because they were regarded as being hard to digest. Lamb's meat was to be refused "becauſe of hiſ exceeding moyſture".

Eggs were passable food during the winter months but not good in summer. Freshwater fish were fine, so "Perche, Barble, Gudgin, Loche, Coole, Troute, and Pyke" were good, and sea fish such as "Gilthed, Turbet, Sole, Rochet, Gurnard, Lapster [lobster], *Crabbe, Praunes, Shrimpes and Whiting,*" were fine providing they were eaten with vinegar.

On the vegetable front Kellwaye lists "*Parfly, Lettis, Sorrell, Endiue, Succorie, Sperage, Hop-buds, Burnet, Borrage, Buglos, Time, Myntes, Ysop*" as recommended for summer pottages, but in winter, "*Balme, Bittaine, Time, Marigolde, Ifoppe, Marioram, Mynts and Rue.*" On the salad front he says: "*For your sallets, take Pimpernell, Purflane, Myntes, Sorrell, Horehounde, Yong cole, Hop-buds, Sperage, Time, Tops of fennell, Tarregon, Lettis, And watercresses are good.*" Pimpernell here is *Poterium sanguisorba*, or salad burnet, and *bittaine* is probably betony.

Similar dietary recommendations appear on the Continent too, and are reflected in works such as *Difesa Contra la Peste* [Marcello Squarcialupi, 1565] and *Instruttione Sopra la Peste* [Michele Mercanti, 1576] as well as many others. Some authors and medical authorities may have occasionally changed the list, but the core dietary advice seems to be quite uniform across Europe.

In the 1649 English translation of Ambroise Paré's works [*The Workes of that Famous Chirurgion Ambrose Parey*], which originates from roughly the middle of the previous century, we find a few extra additions to the *off-limits* plague menu: "*For the Plague often follow's the drinking of dead and muftie wines, muddie and ftanding water, which receiv* [sic] *the finks and filth of a Citie; and fruits and pulf* [sic] *eaten without difcretion in fcarcitie of other corn, as Peaf, Beans, Lentils, Vetches, Acorns, the roots of Fern, and Grafs made into bread. For fuch meats obftruct, heap up ill humors in the bodie, and weaken the ftrength of the faculties, from whence proceed's a putrefaction of humors, and in that putrefaction a preparation and difpofition to receiv, conceiv, and bring forth the feeds of the Plague...*" Again, it is the diet and its' effect on the *Humours* that the concern appears to focus upon.

Paré's inclusion of avoiding drinking from muddy and standing water would probably have been a widely accepted or understood norm, or one would have thought so. If enough villagers, and even village idiots, became ill after drinking from a particular well or stream surely the penny must have dropped for most people that the water source was not good. That said, there are examples today where NGOs (non government organizations) working in the developing world still have to drum home the message of drinking safe water, so who knows if the illiterate peasantry of old fully recognized the problems. On the other hand when you look at Bullein's general health treatise *Bulwarke of Defence* [1579], written around the same time as the original Paré text, it would suggest that there was full awareness of water contamination: "*Water is one of the foure Elements, more lighter then* [sic] *Earthe, heauier then* [sic] *fyer and aire. But this Water whiche is here amonges us in Riuers, Pondes, Springes, Fluds, and Seas, is no pure water, for it is mingled with fundrye Ayres, Corruptions, Groffnes and foftnes. Not withftandinge in all oure Meates and Drynkes Water is ufed, and amongefte all liuinge Creatures cannot be forborne neyther of Manne, Beaſte, Fiſhe, Foule nor Herbe. &c. for all haue neede of water...*" Those few words 'no pure water' would suggest that there was full awareness of potential water contamination.

The list of meats identified as good or bad in Thomas Thayre's plague diet in *Excellent Treatise* [1625] is very similar to Kellwaye and his predecessors going back to Hippocrates, but there are a few slight variations:

"*Fiſhes from freſh Riuers is very good if eaten with vinegar, and good ſauce, they cool the blood well.*

Let your drinke be ſmall beere, and well brewed, and ſometimes a cup of white Wine mixed with water for hot complexions with Borrage and Bugloſſe, but eſchew all hot and ſweete wines.

Hearbes that bee good to be vſed, Sorrell, Endiue, Succory, Borrage, Bugloſſe, Parſely, Mary-golds, Time, Marierom, Betony, Scabious, Iſope, Mints, Purſlane, Pimpernell, Rue, Angelica, Carduus benedictus, Lettuce.

Make your ſauce with Cytrin, Lymons, Orenge, Sorrell, Vinegar, Maces, Saffron, Barberies, and ſuch like.

Raw, and young fruite is hurtfull, ſo is Garlicke, Onions, Leekes, Radiſh, Rocket, Muſtard, Pepper and hot ſpices, and all hot wines, and all theſe are hurtfull, and ſo are all ſweetmeates: let your dyet be cooling and drying."

That final mention of garlic-type ingredients being problematic is mirrored in Browne's *Practical Treatise* [1720] where, referring to the '*Sweating Sickeness'* which appeared in earlier times, he says of the Continentals that: "*... the Reaſon why theſe People abound ſo much with this Diſeaſe is, becauſe they delight in the hot acrid Diet of Onions, Garlick, Sallery, &c. Beſide, the Air is more intemperate and unequal with them than us, from whence the Tone of the Muſcles and Fibres become relaxed, Perſpiration is remitted, whereby their Bodies readily receive Epidemical Diſeaſes, bloody Fluxes, &c."*

Elsewhere in his book Browne has an update on aspects of the Hippocrates diet: "*In Times of Infection, avoid all windy Things that are produced from the Garden, and ſuch, as are ſubject to ſudden Putrefaction, as all Kinds of Pulſe, Cabbage, Colly-flowers, Sprouts, Melons, Cucumbers; &c. as alſo a great many Summer Fruits had better be refrain'd, ſaving Cherries, Currans, Strawberries, Raſpers, Mulberries, Quinces, and Pomgranates, which may be eaten moderately with good Effect.*

Fiſh in general is bad, and ſhould be ſeldom eaten; but the beſt are Soale, Mullet, Plaiſe, Flounder, Trout, Gudgeon, Lobſter, Cray fiſh and Shrimps: Pond-Fiſh are not good, neither Fiſh-Sauces made of hot Spices and Anchovy; but the moſt plain is the beſt, as freſh melted Butter, with Lemon, Orange, or Vinegar."

Habit des Medécins, et aütres personnes qui visitent les Pestiferés, Il est de marroquin de levant, le masque a les yeux de cristal, et un long néz rempli de parfums

The image of the *Plague Doctor* is one of the most enduring plague icons from the 17th century, yet the protective clothing worn has its' origins in Venice and a doctor by the name of Charles de Lorme; physician to the French king Louis 13th, and who also attended some of the Medici family. The cloak was made of thick leather while the beaky face mask was filled with aromatic herbs intended to purify the air, and no doubt the stench of death in the streets and houses. Lorme's innovative design was copied throughout Europe.

PREVENTION and SELF-PRESERVATION

While good diet might help keep your body system in a healthy state and better able to resist infection there were other recommended means of preventative action and self-preservation. The most obvious of these were 'preventative' potions, many being covered in the plants section of this book, while we have touched on the notion of eating healthy food. On the public, rather than the personal, individual, front cities throughout the centuries posted regulations during plague outbreaks to prevent movement of people and animals, and confined members of plague affected households to their dwellings. Perhaps the best known folklore legacy of the 1665 outbreak in London was the painting of a cross on the door of infected households.

That same year the Royal College of Physicians of London produced a pamphlet offering their own advice, as opposed to official regulations. For the 'Prevention of diſperſing the Contagion amongst Perſons' they suggested: "It is adviſable, That all needleſs Concourſes of People be prohibited; That the Poor be relieved and ſet at work, and Beggers not ſuffered to go about; That all ſale of corrupt Proviſion for Food be reſtrained; That Streets and Houſes be as diligently and carefully as may be, kept clean; the Streets waſhed and cooled as much as may be, by the plentiful running of Conduits and Water otherwiſe procured." The College also recommended that no clothes from infected houses could be sold for six months after the infection had ceased in a household, and that the clothing should be 'aired and fumed'; actions which are mirrored in many texts over the centuries.

Nearly seventy years before the famous London pandemic of 1665 Kellwaye [1593] had suggested that the following items should be among the '... orders magiſtrates and rulers of Citties and townes ſhoulde cauſe to be obſerved':

1. Firſt, to command that no ſtincking doonghils be ſuffered neere the Cittie.
2. Euery euening and morning in whot [sic] weather to cauſe colde water to be caſt in the ſtreetes, eſpecially where the infection is, and euery day to cauſe the ſtreets to be kept cleane and ſweete, and clenſed from all filthie thinges which lye in the ſame.
3. And whereas the infection is entered, there to cauſe fires to be made in the ſtreetes euery morning and euening, and if ſome frankincenſe, pitche, or ſome other ſweete thing be burnt therein, it wilbe [sic] much the better.
4. Suffer not any dogs, cattes, or pigs, to runne about the ſtreetes, for they are very dangerous, and apt to carry the infection from place to place.
5. Command that the excrements and filthy things which are voyded from the infected places, be not caſt into the ſtreetes or ryvers whch are dayly in uſe to make drinke, or dreſſe meate.
6. That no Chirurgions, or barbers, which uſe to let bloud, doe caſt the ſame into the ſtreetes or ryvers.
7. That no vautes or preuies, be then empted for it is a moſt dangerous thing.

8. That all Inholders, doe euery day make cleane their ſtables, and cauſe the doong and filth therein to be carryed away out of the Cittie: for by ſuffering it in their houſes, as ſome doe uſed to doe, a whole weeke or fortnight, it doth ſo putrifie, that when it is remoued, there is ſuch a ſtincking ſauour and unwholſome ſmel, as is able to enfect the whole ſtreete where it is.

9. To commande that no hempe or flaxe be kept in water neere the Citte or towne, for that will cauſe a very dangerous and infectious ſauour.

10. The haue a ſpeciall care that good and wholſome victuals and Corne, be ſolde in the markets, and ſo to prouide that no want thereof be in the Cittie, and for ſuch as haue not wherewithall to buy neceſſary foode, that there to extende their charitable and godly deuotion: for there is nothing that will more encreaſe the plague, then [sic] want & ſcarſitie of neceſſary foode.

11. To commande that all thoſe which doe viſit and attende the ſicke, as alſo all thoſe which haue the ſicknes on them, and doe walke abroad: that they do carry ſome thing in their handes, thereby to be knowne from other people.

Knowing what we do about *Yersinia* these days, some details of the above for controlling plague are quite obvious, but there are interesting points in the passage that reflect matters already touched upon in previous texts so far. For example, the mention of keeping the city supplied with food hearkens back to the notion of plague following on from times of famine. It is also quite clear to see the then current thinking of smells and noxious fumes having some part in the transmission of the disease. The further provision of identifying anyone in the street connected with management of the plague outbreak would also be a valid safety measure since many physicians, nurses and morticians became both carriers and casualties of the disease.

Around the same period as Kellwaye's thoughts William Bullein was suggesting the following defensive plague actions to the readers of his *Dialogue Against the Fever Pestilence* [1579]: *"Surely I wil [sic] declare thee the beste defence that I can; I will hide nothyng. First of all, let all men, women, and children auoide out of the euill ayre into a good soyle, and then, accordyng to their age, strength of nature, and complexion, let euery one of them with some good medicine drawe from the bodie superfluous moysture, and diminish humour, hotte and drie, and vse the regiment of diet to driyng [sic], sharped with vineger or tart thynges, and lesser meates; not so much wine as they haue vsed in custome; neither Potage, Milke, vnripe fruites, hotte Spices, Dates, or Honie, or sweete meates, wine with Suger, are not tollerable; no anger or perturbations of the mynd, specially the passion called feare, for that doth drawe the spirites and blood inwards to the hart [sic], and is a very meane to receiue this plague; neither vse actes venerous, nor bathyng, either with Fume, stoue or warme water—they all doe open the pores of the bodie; neither quaffyng or muche drykyng. Euen so thirste or drinesse is not tollerable, or immoderate exercise or labour, specially after meate. Music is good in this case, and pleasaunt tales, and to haue the meates well sauced with cleane sharpe vineger. Forget not to keepe the chamber and clothyng cleane, no Priuies at hand, a softe fire with perfumes in the mornyng. Shifte the lodging often time, and close in the Southeaste windowes, specially in the tyme of mistes, cloudes, and windes; And vse to smell vpon some pleasaunt perfume, And to bee letten bloud a little at once, and to take Pilles, contra pestem: that is a good preseruative against the plague.*"

The anonymous compiler of *Some Observations Concerning the Plague* [1721] has a précis of the health regime that the Dutch physician Diemerbroek claims to have used for his personal self-preservation during the plague outbreaks in Nijmegen during 1635 and 1637. Diemerbroek's work on the plague was widely quoted among medical writers during 17th and first part of the 18th century so it is interesting to see that this précis appears in a booklet aimed at the lay-person or concerned citizen, and perhaps suggests that members of the public would have followed Diemerbroek's advice had there ever been another Black Death or Plague of London:

"*The Sum of what he says, I have endeavoured to contract as follows: He tells us, that he avoided as much as he could all vehement Pertubations of Mind: That he lived intrepidly, or without Fear: That it was the same Thing to him whether he visited the Sick of the* Plague *or of any other Distemper, and that he as readily served the Poor* gratis, *as the Rich for a* Reward; *He adds, that if at any time he found himself some-what shocked, (which in that doleful Season, wherein there was scarce an House in the whole City that, escaped the* Contagion, *must needs happen now and then) in such a Case he refreshed his Spirits with three or four Draughts of* Wine; *That being frequently disturbed in the Night, and much tired in the* Day *with walking up and down from* Patient *to* Patient, *he was forced to Sleep an Hour after Dinner, when he could best spare the time, though he disswaded others, who were under no such necessity, from sleeping in the Day: That as to Diet, he used Meats of the most easy Digestion, avoiding* Swine's Flesh, Herrings, *and the like, which he had found hurtful to him: That his* Drink *was ordinary* Ale *and Small* White Wine, *of which he sometimes drank to Chearfulness, but never to Excess; That he kept his Body open, but not too loose, only so as to have* One *or* Two *Stools in a Day: That* Once *or* Twice *in a week at Bed-time he swallowed* One *or* Two *of his Anti-pestilential Pills... That beginning to visit the Sick between* Four *and* Five *in the Morning, he could then take Nothing, his Stomach perfectly loathing both Meat, Drink, and Medicine, so that he was constrained (though against his own Judgment) to go forth fasting, and could do no more than (after Committing himself to* God *by pious Prayers) to chew some Grains of the lesser* Cardamon; *That about Six a Clock in the Morning; he took a little* Treacle *or* Diascordium, *or eat a little* Candy'd Orange Pill, *and very frequently three or four Bits of* Candy'd Elecampane: *About* Eight *he breakfasted upon a Piece of* Bread *with* Butter *and* Green-cheese *made of Sheeps Milk, drinking a Glass of Ale, and now and then (but not daily) he took a Draught of* Wormwood Wine *about* Nine: *At* Ten *he smoaked a Pipe of* Tobacco, *and after Dinner* Two *or* Three, *and the like after Supper, and sometimes* Two *or* Three *more between Meals; and if at any Time he found himself affected with the* Ill Smell *of the Sick, he presently had*

Recourſe to the ſame Remedy, *which he ſays, he found by his own Experience, as well as always thought to be the moſt effectual* Preſervative, *ſo that the* Tobacco be of the beſt Sort. *He adds, that upon the ceaſing of the* Plague *he left off* Smoaking, *not willing to accuſtom himſelf to it, left he ſhould turn its laudable Uſe into a deteſtable Abuſe."*

The use of tobacco in herbal medicine had occurred within a matter of years of its appearance in Europe and, indeed, there is at least one 17th century book in English wholly dedicated to the herbal and medicinal properties of tobacco. So it is interesting to see tobacco appearing to such a large extent in Diemer-broek's health regime above, but also somewhat amusing that he appears to regard the plant as having pernicious effects outside of medicinal use. Among the herbal plants used are candied orange peel and elecampane, and there is more about their use in the later part of this book. The use of wine in both treatment regimes and as a preservative was also quite common, Joseph Browne [1720] commenting around the same time as the booklet above was published: "... *fine Mead is of excellent Uſe, and good Wine is an Antidote againſt all Poiſons; but beware Exceſs,"* though Browne makes reference to the work of Celsus for this advice.

Should someone find themselves passing through an area or place where the air was thought to be contagious or contaminated then one option was for the individual to undergo preventative sweating, using a sudorific or diaphoretic rather than a bath. In the passage below, from the posthumous Thomas Willis text [1691], the advice is to stay indoors post-sweat so that the skin pores can close and so prevent infection entering by that means: "*Alſo for thoſe that live in an infected Air, that there is ſuſpicion that they may daily take in ſome peſtiferous Vapours, which fermenting with the Blood and Humours, may infenſibly at laſt break out in the Plague; it may not be amiſs, once or twice a week, to take pretty large Sweats in their Beds: And this to be done, eſpecially if the Party has had any occaſion whereby he may ſuſpect himſelf to have been more open to infection, or that he has taken any: After ſuch Sweat he ſhould keep his Chamber the forepart of the day, till the Pores are reduced to be as they were before."*

Another small paragraph in Willis' text seems a little at variance with Diemerbroek on the matter of going out in the morning without eating. Other texts too mostly recommended not going out on an empty stomach: "*Let none go Abroad Faſting, but every Body eat according to their Cuſtom and Circumſtances, as Bread and Butter with Rue, Sage, Sorrel, &c. or a Toaſt ſop'd in Wine or Metheglin. The Cuſtom that prevails now may be of excellent Uſe, that is, to Break-faſt upon Coffee, Bohea Tea, or Chocolate, with Bread and Butter."*

The 18th century German surgeon Lorenz Heister [1750] called upon fellow physicians to eat before they went to work on their plague rounds: "... *there are feveral human Cautions and Obfervations neceffary to be regarded by the Phyfician and Surgeon; the chief of which are, that they fhould never go fafting to vifit a Patient fick of any contagious Difeafe, and much more of the Plague; but they fhould always eat fomething, and drink fome ftrong Liquor before-hand, in order to defend themfelves from the peftilential Contagion and infected Air: Some Phyficians therefore always eat Bread and Butter, and drink a Draught of Spanifh or Wormwood Wine, or fome other ftrong Wine, before they offer to fet a Foot in the Patient's Houfe..."*

The personal use of vinegar as a form of disinfectant, and odoriferous herbs or compounds, was a widely held belief. Bullein's [1579] recommendation of opening the windows and using vinegar in the bed-chamber was typical: *"Drawe the Curtaines, open the luket of the windowe, set Sallowes about the bed besprinkled with Vinegar and rose water."*

Similar advice is given by Joseph Browne [1720] more than a hundred years later: *"Herbs, Rufhes, and Boughs that are neceffary to be difpofed about the Houfe and Bed-chamber, which yield refrefhing Scents, and contribute much towards purifying the Air, and refifting the Infection. Of this kind are all forts of Rufhes and Water Flags, Mint, Balm, Camomil Graffs, Hyffop, Thyme, Penny-royal, Rue, Wormwood, Southern-wood, Tanfy Coftmary, Lime-tree, Oak, Beech, Walnut, Poplar, Afh, Willow, &c."*

In a reference to the works of the Italian poet-physician Frascastorius – who lived in the 15th-16th century period, and was partly responsible for propagating the theory of contagion and germs being the cause of disease – Browne continues: *"Likewife Linen Cloths may, be dip'd in Vinegar and any fweet fcented Water, as Tanfey, Angelica, &c. and thefe Cloths to be hung upon the Walls, or upon Cords in Bed-chambers, being every Day wafh'd and dip'd afrefh, as before; by which means they will attract and imbibe the Virus which floats in peftilential Airs; efpecially, if after wafhing and dipping, as before, the faid Cloths or Sheets hang a little before a brisk Coal Fire till they fmoak, their Power of Attraction will be much the greater..."*

In some advice to his fellow medics as they went about their daily business tending to the sick Heister [1750] suggested that: *"... in order to keep off or correct the peftilential Effluvia, it will not be improper frequently to hold a Sponge to the Nofe which has been firft wetted with fimple Vinegar, or that wherein Rue or Lavender has been infufed..."*

The use of vinegar was not uncommon even a hundred years later, the English physician Richard Pearson [1813] penning an enlightened plague treatise for the time for the use of his fellow practitioners, saying:

"Those who, from their situation or business, cannot always avoid actual contact of infected persons or infected things, should pay particular attention to ablution, which should be repeated every time they may have touched infected clothes, furniture, or persons. For the purpose of ablution, cold water mixed with vinegar, or vinegar

alone, may be used. Not merely the hands, but the arms, neck, and face, should be well washed with it. Thus if any of the contagious particles should adhere to the skin, they will be removed from it before there is sufficient time for their absorption.

It is obvious that cleanliness is as necessary in regard to linen and clothes, as it is in regard to the body itself; they should therefore be frequently changed and washed."

In another part of his book Pearson also makes an interesting suggestion for the use of salt water as a preventative measure: "Where access to the sea cannot be had, nor an artificial salt water bath on a sufficiently large scale, I propose, as a convenient substitute, that the shirt be dipped every morning in a saturated solution of common salt in cold water, and that after having been gently wrung out, it be put on wet and cold. This will produce all the tonic effects of sea-bathing, and may in all cases be prescribed where that remedy is proper. Immediately after putting on the wet salted shirt, persons should not sit still, but keep the body in motion. This preventive measure is particularly adapted to the warm season of the year, at which time, in Europe, the plague always rages with the greatest fury."

The 19th century English military physician Tully, mentioned elsewhere here, also refers to a regime of preventative sea-water bathing for inhabitants on the Ionian island of Cephalonia when he was stationed there. There was a suspicion that those involved may have come into contact with plague carriers and he tells us in his book [1821] that: "... the inhabitants of each tent [ED – where they were quarantined] were marched to and from the sea-side, in separate bodies, by health guards." The actual immersion regime he describes as: "... the whole having been plunged daily into the sea, without regard to age or sex, and all of their susceptible effects having also been daily immersed in sea water, for the space of two hours, and subsequently exposed to the heat of the sun, the thermo-meter in the shade on the beach standing at noon at 88°." It would be interesting to know whether Tully acted upon Pearson's previous salt-water immersion suggestion or whether this preventative method was common among English or other physicians.

Perhaps one of the most memorable impressions one gets from reading about the Great Plague of London, and others, is that of fumigating both private homes and the streets with smoke, often acrid (as when sulphur was burned) though sometimes scented with odoriferous plants and other aromatic ingredients. In the very oldest plague and pestilence texts fumigant ingredients were some-what limited in scope, substances such as juniper and bay leaves being commonly used. As the centuries progress it seems that the number of aromatics employed as fumigants increased, and by the time of the 1720 Marseilles plague outbreak even poisonous mercurial compounds were in the fumigant frame.

Andrew Boorde in his Compendyous Regiment or Dyetry of Health [1542] which encompasses domestic health and hygiene, suggests the following daily fumigation routine for his readers: *"... wherfore in such infectious tyme it is good for euery man that wyl not flye from the contagyous ayre, to vse dayly—specyally in the mornynge and euenyng—to burne Iuneper, or Rosemary, or Rysshes, or Baye leues, or Maierome, or Frankennce [sic], [or] bengauyn. Or els make this powder: Take of storax calamyte half an vnce, of frankensence an vnce, of the wodde of Aloes the weyghte of .vi d [drachms]; myxe al these togyther; Than cast half a sponefoll of this in a chaffyng - dysshe of coles, And set it to fume abrode in the chambers, & the hall, and other howses. And you wyll put to this powder a lytell Lapdanum, it is so moche the better. Or els make a pomemaunder vnder this manor. Take of Lapdanum .iii. drammes, of the wodde of Aloes one drame, of amber of grece .ii. drames and a half; of nutmegges, of storax calamite, of eche a drame and a half; confect all these together with Rose-water, & make a ball. And this aforesayd Pomemaunder doth not onely expell contagyous ayre, but also it doth comforte the brayne, as Barthelmew of Montagnaue sayth, & other modernall doctors doth afferme the same..."*

Across the English Channel Royet, in *Excellent Traicte De La Peste* [1583] had his own suggestions on the aromatic incenses to burn in rooms and for fumigating clothes: *"... on les perfumera de chofes aromatiques, comme d'encens, myrrhe, benioin, ladanum, ftyrax, rofes, feuilles de myrthe, lauande, rofmarin, fauge, bafilic, farriete, ferpolet, mariolaine, geneft, pommes de pin, petites pieces de bois de pin, de geneure, & fa graine, clous de girofle, oyfelets de Cypre, & autres femblables chofes odoriferantes. Et de cefte mefme fumee faut parfumer les habillemens."*

In the 18th century the plant species in Joseph Browne's [1720] recommendation for aromatic herb purifiers, as opposed to heated fumigants, is much more Anglo-centric and draws less upon plants from warmer climes: *"Herbs, Rufhes, and Boughs that are neceffary to be difpofed about the Houfe and Bed-chamber, which yield refrefhing Scents, and contribute much towards purifying the Air, and refifting the Infection. Of this kind are all forts of Rufhes and Water Flags, Mint, Balm, Camomil Grafs, Hyffop, Thyme, Penny-royal, Ruw, Wormwood, Southern-wood, Tanfy, Coftmary, Lime-tree, Oak, Beech, Walnut, Poplar, Afh, Willow, &c."*

Thomas Thayre [1625] talks about burning oak wood as a street fumigant with some juniper sticks and perfume added to make the fumes less acrid, and had the following recipe for daily household use: *"R. Storax, Calamint, Labdanum, Cypreffe-wood, Frankencenfe, Beniamin, of each of them halfe an ounce; red Rofe-leaues dried, yellow Sander, of each two drams; Cinamon, Cloues, wood of Aloes, of each of them one dram; flowers of Nenuphar one dram; liquid Storax half an ounce, gum Dragagant two drams, and muske fix Graines, Withy cole three ounces, Rofe-Water as much as will fuffice to make it up in Trochis."* Withy cole is willow charcoal, which would have been the combustible material here.

Thomas Coghan, in a similar domestic health and hygiene work of the same period, *Haven of Health* [1636], covers some of the same fumigant ingredients but provides other alternatives for poor people who would have not been able to afford items such as cloves or frankincense: "*And it fhall be good, fpecially at night and in the morning, to perfume your houfe or chamber with frankinfcenfe, or Iuniper, or ftorax Calamita, or Ladanum, or if you will not be at coft, with dryed Rofemary, or as poore folkes ufe to doe in great townes, with rufhes or broome, or hey layed upon a chafing difh and coales, and the windowes and dores being clofe fhut up for the time. Or to heate a bricke or flate in the fire, and when it is hot, to take it out, and poure vineger* [sic] *upon it, and to receive the fume with open mouth. But among all things that purifie the ayre, either within the houfe or without, none is better than fire: for fire by nature doth confume corruption...*"

In the period of the 1665 Great Plague the Royal College of Physicians of London put out their own suggestions in a pamphlet:

"*Brimftone burnt plentifully in any room or place, though ill to be endured for the prefent, may effectually correct the Air for the future.*

Vapours from Vineger [sic] *exhaled in any room, may have the like efficacy; efpecially after it hath been impregnated, by infufing or fteeping in it any one or more of thefe Ingredients; Wormwood, Angelica, Mafterwort, Bay leaves, Rofemary, Rue, Sage, Scordium, or Water-germander, Valerium* [sic]*, or Setwall-root, zedoarie, Camphire.*"

The first line of the text above highlights one of the problems of burning sulphur in air, particularly in closed rooms, that of sulphur dioxide production. On a purely practical level sulphur dioxide might possibly have done wonders and killed off any flea vectors which could transmit plague from rats, though that does not appear to have been the intention. Rather, burning sulphur produces an irritating, pungent, gaseous compound (the SO_2 mentioned) that can produce feelings of chest tightness and coughing, with asthma sufferers being considered particularly sensitive to the gas. In a similar toxic vein the author of *Avis de Precaution* [1721] suggests fumigating houses with cinnabar, orpiment, arsenic, antimony and salt-petre, but notes that these are '*pernicieufe aux perfonnes*' and that once lit you should retire from the room promptly and close the door behind you. Probably a wise move since cinnabar and orpiment are mercury and arsenic sulphides respectively, while antimony, a chemical element in its own right, will cause a similar type of poisoning.

For domestic purposes fumigrant ingredients could be burnt on a chaffing-dish of coals, while Joseph Browne recommended fumigant fires should be made in small stoves that could be carried from room to room, rather than fires lit in a fireplace where the perfumed smoke would simply escape up the chimney rather than do its job of purification. He goes on to give a number of recipes, four of which follow below and give an insight into the make-up of the fumigants and means of application:

"Drs. *Butler* and *Atkin's* Method to Fume the Houſe.

Take a Quantity of ſtrong Vinegar, and put a little Roſe-water, and a good deal of Roſemary thereto; put them all in a Baſon; then take Five or Six Flint Stones heated red hot, caſt them into the Vinegar, and let the Fumes be convey'd over all the Houſe. Likewiſe ſcent it with Rue, Gentian, Angelica, Juniper, Lavender, Roſemary, Mint dry'd and burnt over the Fire; or elſe firſt ſteep'd in Wine Vinegar, and then burnt. Or ſcent the Houſe and all the Furniture with quick Lime ſlack'd in Vinegar. Or burn Pitch, Tar, Roſin *or* Frankincenſe."

"Another way to Perfume the Houſe or Bed-chamber.

Take Wood of Aloes 3ij. Cloves, 3ſs. Juniper and Bay Berries, of each 3iſs. Sage, Roſemary and Marjoram, of each two handfuls, make a groſs Powder, which burn either in your Bed-chamber, or elſe-where in the Houſe."

"Dr. Goddard's *Liquid Fumes or Vapours, were made of Orange Peels, Cloves, Cinamon, Aloes Wood, Citrine Sanders, Enula Campana Roots, Red Roſe Leaves, and a little Camphore, fumed in Vinegar of Roſes, in the Bed-chamber, or by the Patient's Bed-ſide, by means of a Chaffing Diſh of Coals, or a Lamp burning with Spirit of Wine camphorated, or Spirit of Lavender for Perſons of Diſtinction.*"

"Dr. *Gliſſon*, Sir *Tho. Millington*, Dr. *Charlton*, *and other learned Phyſicians, I find, in the laſt Plague, recommended Fumigations to be made of* Arſenick 3ij. *and* Sulphur 3ſs. *eſpecially in Places infected; but this was done in the empty Rooms, which were afterwards fumed with* Cloves *groſly pouder'd* ℥i, Roſe-Water ℥iij, Vinegar ℥ij, *burned in a Pan of Coals; or elſe by caſting a Pint of White Wine-Vinegar upon the Fire, in the Midſt of the Bed-Chamber.*"

In the posthumously published work of Thomas Willis [*Plain & Easie Method*, 1691] we find: "*Beſides the ſuppreſſing of Vapours that may increaſe the infection of the* Air *it is to be purg'd of that Malignity it brings with it from other infected places; and this is done by great Fires, which ſhould be continually kept, except the Weather be too hot, and by Fumes of* Sulphur, Nitre, Frankincenſe, Pitch, Roſin, Tar, *and the like, which every day ſhould be burnt in the Room we moſt frequent, alſo before our Doors, and on the tops of our Houſes. Of ſimple Medicines to be us'd for this purpoſe,* Brimſtone *is commended for the beſt that is; 'Tis likely that* Vitriol, *which partakes much of the like acid Spirit, may be very proper; but in regard 'tis not eaſily combuſtible...*" There then follows a fumigant recipe with calcined green vitriol, saltpetre and sulphur which was to be strewn on coals in a chaffing-dish.

The French work *L'Ordre Public Pour la Ville De Lyon*, published in 1670, largely deals with the mechanisms for quaranteening the public, but includes a few preservatives and fumigants, and among them the following *Compoſition du Parfum à des-infecter les maiſons* which has close similarities to some of the previous ones, but here the quantities are for use in public fumigation situations, the ingredients measured in pounds, starting off with an enormous quantity of sulphur: "*Souffre nonante-huit livres, antimoine fept livres, tartre fept livres, poudre fine de chaffe trois livres & demy, carabé vne livre & trois quarts, arfenic trois livres & demy, orpiment vne livre trois quarts, canfre vne livre cinq onces. Eaut bien faire piler, le tout feparément, & ayant fait fondre le fouffre dans vne marmite de fer, les poudres cy-deffus bien mêlées, font mifes dans ledit fouffre peu à peu; ce fait l'on a vne pierre cavée de la grandeur que l'on veut faire les pains, & ayant mis du papier fur ladite pierre; l'on jette la compofition deffus, don't il fe fait plufieurs pains.*"

The usefulness of burning fires in streets was questioned by some. Blackmore [1722], in the text below for example, even suggests that by heating the air (so expanding it) allows it to draw in infected air, basing his argument on the fact that plague appeared more often in summer: "*As to the making of great Fires to hinder the fpreading of Infection, it feems much more detrimental than ufeful. For it is evident that Heat does not purify the Air, and free it from the Peftilential Particles that inhabit there; but on the contrary, by opening and inlarging its Pores, it gives an eafy Admiffion to fuch Exhalations: And this is evidently confirm'd by this undeniable Obfervation, That the Peftilence advances, and fpreads its contagious Fury in the Summer-Season, but is fuppref's'd or chek'd in its Progrefs, while the Frofts and rigorous Cold of Winter, ftreighten and contract the Pores or Vacancies of the Air, by drawing its Parts clofer together, by which Means the noxious Seeds of the Plague cannot enter and be receiv'd in fo great plenty.*"

While burning sulphur and saltpetre on fires were popular activities, another option that seemed to have popularity from time to time was firing cannons. Gunpowder is, of course, composed of sulphur, saltpetre and charcoal, so one can see the connection in the old logic; don't burn it, detonate it instead, and spread toxic fumes that way. Texts, however, do seem to suggest that it was also believed that the reverberatory effects of cannon fire as it rattled through streets had microbe killing capacity.

Regarding the '*difcharging of peeces*' Bradwell [1636] says: "*... for Gunpowder is exceeding drying by reafon of the Salt-peeter and Sulphur with which it is made, and by the crackes that it gives, the Ayre is forcibly fhaken and attenuated, and fo opened to let in the purification, which is immediately made by the fire that goes along with it.*"

The usefulness of firing of cannon and lighting fires appears to have been a contentious issue as suggested in this next passage from Joseph Browne [1720]. Richard Mead and Nathaniel Hodges were two key observers and writers on the plague during the 17th-18th century period and had questioned the point of fires and firing cannons, but obviously they did not convince Dr. Browne who also believed the tarry chemicals in coal had bug-busting potential. Indeed, coal tar is a source of phenol which has antiseptic and germicidal properties and is one of the recommended types of disinfectant to be used today during plague outbreaks: "*Whatever Dr. Mead has urged againſt Fires being made in the Streets, built upon Dr. Hodges de Peſte, the Experience of all Ages, and all Authorities, are againſt him; but had that not been ſo, there is the ſtrongeſt Reaſon in Nature for Fire in peſtilential Times, eſpecially Coal-Fires, by Reaſon of the great Quantities of Amber contain'd in that Body, as our Modern Chymiſts know to their great Gain, and Abundance of Bituminous Earth, both which are great Enemies to Contagion, and Deſtroyers of the malignant Virus, whether it be a verminous, or any other kind of corrupted Exhalation: for as Fires diſſipate the collected Atoms in peſtilential Airs, by rarifying and attenuating the groſſer Particles of the Atmoſphere, within the Compaſs of their activity; ſo the Experience, both of Soldiers and Seamen, will juſtify, that firing of Guns, eſpecially Cannon, will purify the Air, both by Concuſſion, as well as by its conſtituent Parts of Sulphur and Niter...*"

Gunpowder itself could also be burnt as a fumigant as De Mertens [1799] briefly mentions in his account of the 1771 Moscow plague, before outlining a mixture that seems to have had some success: "*The houſes and rooms of perſons infected with the plague are purified by firing gunpowder in them. At Moſcow we employed with ſucceſs a powder, called* antipeſtilential, *of which ſulphur and nitre formed the baſis; ſome bran and other vegetable ſubſtances, ſuch as abrotanum, juniper-berries, &c. together with certain reſins, were added; but in my opinion theſe reſins are totally uſeleſs, and only increaſe the expence.*"

De Mertens also warns of the dangers to personal health when using these fumigants: "*This vapour is hurtful to the lungs, and produces ſuffocation; hence the perſon who throws the powder upon the burning coals ſhould get out of the room as faſt as poſſible. This process is repeated three or four times in the ſpace of twenty-four hours for ſeveral days together; after which the doors and windows are thrown open.*"

The same 1799 English translation also lists the composition of the following fumigant powders published by Moscow's Council of Health at the time of the 1771 outbreak, each mixture having a specific, rather than generalized, application:

"The ſtrong antipeſtilential powder *conſiſted of juniper tops (cut ſmall,) guaiacum ſhavings, juniper berries, bran, of each 6 ℔, nitre 8 ℔, ſulphur 6 ℔, myrrh 2 ℔.*

The weaker antipeſtilential powder *conſiſted of the herb abrotanum 6 ℔, juniper tops 4 ℔, juniper berries 3 ℔, nitre 4 ℔, ſulphur 2½ ℔, myrrh 1½ ℔.*

The odoriferous antipeſtilential powder conſiſted *of calamus aromaticus 3 ℔, frankincenſe 2 ℔, amber 1 ℔, ſtorax and dried roſes, of each ½ ℔, myrrh 1 ℔, nitre 1 ℔ 8 oz. ſulphur 4 oz.*

Of theſe powders, the firſt was employed to fumigate the houſes and goods of the infected, ſuch as woollens, furs, &c.; the ſecond, for fumigating houſes only ſuſpected, and more delicate articles, which would have been ſpoiled by the firſt; the laſt was employed (by way of prevention) in inhabited houſes."

While the resounding, reverberating, clatter of cannon fire around town was likely to do little to attenuate a plague outbreak neither was bell ringing, another of the imaginary preventative public measures. The regular peels of funeral bells would have been a common occurrence during plague, and there were also warning bells rung for things like curfews. However, there was one school of thought which believed that the sustained sonorous output from church bells could have similar effects to the ear-battering of cannon fire. More of a post-Medieval preventative, by the time of Scarborough's *Discourse* [1722] the likely usefulness of bell ringing was somewhat in doubt: "*As to the ringing of Bells to purge and refine the Air, and render it wholeſome by the Diſperſion of the Peſtilential Matter lodg'd within it, it will eaſily be ſeen that this Expedient can avail little; for if the conſtant ringing of Bells could, by the Agitation of the Air round about, diſſipate and conſume the malignant Impurities that adhere to it, this Effect would more certainly be procur'd by the ſtronger Force of high Winds and Tempeſts, which, however are ſo far from purging the Air from ſuch deſtructive Vapours, that they carry them abroad and ſpread the Contagion farther, without abating their Fury in the infected Places...*"

The final aspect of self-preservation from a plague outbreak was that of fleeing the area, and down the centuries this is a common thread throughout plague texts. While some writers advocated flight, others rightly identified that the plague could be spread through travel by plague-carrying individuals, and an occasional few reflect on cowardly physicians, some clergy, and well-healed intelligentsia who left the poor behind unattended while saving their own skins.

In the 17th century Thomas Archer [1625] reckoned flight was the cheapest preservative: "*... wee muſt flie away ſpeedily, and wee muſt go farre off, and returne ſlowly, it is good for thoſe that can conueniently ſo doe.*"

Thomas Coghan in *Haven of Health* [1636] put flight as his primary means of self-preservation: "*The firſt way then of preſervation from the Plague, is, with ſpeed to goe farre from the place infected, and there to remaine untill all the infection be paſt.*"

The character Mendicus in William Bullein's *Dialogue Against the Fever Pestilence* [1579], paints the following vivid picture of the exodus of people from populated areas: "*... in the Citie, which doe feare the Pestilence. I met with wagones, Cartes, & Horses full loden with yong barnes, for fear of the blacke Pestilence, with their boxes of Medicens and sweete perfumes. O God, how fast did thei run by hundredes, and were afraied of eche other for feare of smityng.*"

Around the same period Andrew Boorde [1542] seems to put flight among the accepted options available to his readers: "*Whan* [sic] *the Plages of the Pestilence or the swetynge syckenes is in a towne or countree, with vs at Mountpylour, and al other hygh Regyons and countrees that I haue dwelt in, the people doth fle from the contagious and infectious ayre; preseruatyues, with other counceyll of Physycke, notwithstandyng. In lower and other baase* [sic] *countres, howses, the which be infectyd in towne or cytie, be closyd vp, both doores & wyndowes; & the inhabytours shall not come a brode, nother to churche, nor to market, nor to any howse or company, for infectyng other, the whiche be clene without infection."*

Even in the mid-18th century flight was still on the self-preservation agenda, certainly in Continental Europe. In an English edition of a work by the renowned 18th century German surgeon Lorenz Heister [1750] flight appears to be the primary course of action; though he has recommendations for those who cannot flee, and advises physicians that the best way for them to remain healthy was to keep cheerful and confident: "*The beſt and readieſt Defence againſt the Plague ſeems in general to conſiſt in this, that ſuch as are able ſhould remove out of the peſtilential or infected Air into ſome healthy Part of the Country; or wherever they are, they ſhould keep from the Company of ſuch as are already infected, and not meddle with their Cloaths, Bedding, Meat, Drink, or Veſſels, and above all, not to make themſelves over afraid of the Diſeaſe; but let them always keep a chearful and confident Mind, with a proper Diet. But for the Phyſician and Surgeon, whoſe Buſineſs it is to relieve the Sick, and for that Purpoſe muſt enter dangerous Places, it is beſt for them to keep up a courageous Mind, and not be anxiouſly afraid of Diſeaſes, nor even the Plague..."*

Not all plague texts, however, recommended flight and, as some of the writers of passages earlier in this book point out, there were people who were surrounded in their neighbourhood by infected households but who did not fall ill with plague. As we know today plague is generally a contact disease, or problematic where an individual inhales the contaminated respiratory droplets of someone close-by who is infected with the pneumonic form of plague.

Gadbury, in *London's Deliverance* [1665], is somewhat scathing with regard to physicians recommending flight and seriously questions Hippocrates' maxim in time of plague, '*Cito, Longe, Tarde*', or leave quickly, go far away and come back slowly. Although it might not have been regarded as truly heretical Gadbury's questioning of Hippocrates, a cornerstone of medical thinking and action for physicians in past centuries, is unusual. He also takes a swipe at cowards who flee the plague and the general folly of flight in a plague outbreak:

"*The Reaſon why many perſons, ſo willingly* facrifice *to their* Fears, *in flying from their* habitations *in time of a* Peſtilence, proceeds

    1.   *From the* cuſtomary *advice of Phyſicians.*
    2.   *From a principle of* Cowardiſe *in themſelves.*

Firft, Phyficians *in moft* knotty distempers *of a Chronick Nature advife their* Patients *to a change of their* Air, *which is very neceffary, that being one of the fix* non naturales; *for they cannot take too much care of that* Patient, *who groans under the tyranny of Sicknefs. But to prefcribe (unto perfons that* ail nothing*) a neceffity of* removing, *becaufe more perfons then* [sic] *ordinary dye about them, I fee no clear Reafon for.* That *ancient,* but queftionable *Oracle* of *Hippocrates, - Cito, Longe, Tarde;* fhould it with other *Oracles* ceafe, I prefume would be no *injury,* but *advantage* to the human race; fince it may truly be deemed, that the obfervation thereof hath *deftroyed* many more perfons then [sic] it hath *preferved.* How many are there, that by *flying* from dangers, have *fallen* into the middeft of *dangers* ? When men *Plot* to fave themfelves, their contrivances often procure their ruine... Men may *flye* from their *houfes,* their *families,* their *companions friends,* and *relations,* and thereby become examples of *fear* and *terror* to *others;* but they cannot *fly* from *God.* In vain they at all attempt it... *Cowards* hoping to avoid dangers, *rufh* ignorantly into *them.* A *Bullet* may fooner kill *him* that *runs* from the *battle,* then [sic] *him* that ftoutly and refolutely *joyns* therewith; the truly *valiant* often efcape untoucht."

Samuel Pepys, the 17th century diarist, noted in his diary on August 16th, 1665: "*To the Exchange, where I have not been a great while. But, Lord! how sad a sight it is to see the streets empty of people and very few upon the 'Change.*" A couple of weeks later, on August 30th, his diary entry reads: "*... how every body's looks and discourse in the street is of death and nothing else; and few people going up and down, that the town is like a place distressed and forsaken.*" The following day Pepys removed down the River Thames to Woolwich where he was closer to his Admiralty work, but also safer.

Understandably, sectors of populations down the centuries resorted to flight given their limited medical understanding of a dreadful disease that produced terrifying outcomes on most of the occasions where it had appeared throughout history. Add superstition to that lack of understanding and plague was a disaster waiting to happen at any time. Still, had people adhered to the advice given in the 15th century English text, *Regimen Contra Pestilentiam* [1483], then a lot of people might have been safer by staying put in their homes: "*Alfo in tyme of peftilence it is better to abide within the hous* [sic] */ for it is not holfom to goo* [sic] *into the Cyte or towne.*"

FIRST SIGNS and MANIFESTATION

One of the problems that besets any investigation of 'the plague' or historical plague 'events' is that of qualifying the presence of *Yersinia* as the cause. Over the centuries there have been numerous accounts and reports of visitations of plague and pestilential fever across Europe, but they may not all have been plague as we pigeon-hole the disease today. And while we describe five clinical forms of plague, sometimes overlapping as secondary forms, physicians in centuries past often categorized plague fever as simply another form of fever. Furthermore, the word 'pestilence' was often used interchangeably with that of plague, but in different cultures the term pestilence could mean different things, and in translation that difference sometimes gets lost. A simple example is that of Yellow Fever which, in 19th century America, was called 'the pestilence'. So some care needs to be taken when interpreting accounts from the past and across different cultures.

Some of the passages in the following section on the actual diagnosis and treatment of the disease have been left in their long form rather than isolated textual sound-bites, since it is felt the inclusion of surrounding text offers a better insight into the medical reasoning, logic, and thinking that were prevailing at each period.

As discussed previously, contaminated air was a significant player in medical thinking on the plague right down the centuries; and while it was imagined that plague victims acquired the disease simply from breathing day-to-day air present in a plague-laden environment the other links in the infective chain remained elusive. In the following passage, from the posthumously published work of Thomas Willis [1685], the book compiler talks about the *secret and hidden manner* in which the plague could creep up upon the victim, but air is there as part of the infection equation: "The *Plague* may be defcribed after this manner, That *it if an Epidemiouf Difeafe, contagious, very deftructive to Mankind, taking its Rife from a venemous Miafm firft received by the Air, afterward propagated by Contagion, which fetting upon Men after a hidden and fecret manner, caufes Extinctions of the Spirits, Coagulations of the Blood, Syderations and Mortifications of it, and of the folid Parts, and brings the Difeafed in danger of Life, with an Appearance of Pufhes, Bubos, or Carbuncles, and with the addition of other horrible Symptoms.*"

All the same culprits are present in Joseph Browne's [1720] comment below – air, bad diet, infected clothing and possessions, and close proximity to infected people:

"*The Plague is a contagious Venom, or fubtle Poifon, generated within the Body, or communicated from without; being of a malignant epidemical Nature, that fuddenly affaults the vital Strength and Spirits of Mankind, in all Ages of Life; therein putting him in the utmoft Terror and Apprehenfion of Death.*

This Peftilential Virus, *propagated either by the Air, Diet, or infected Goods, Houfes, or Converfation with difeafed Perfons, and, laftly, by Difeafes which are the Caufes of other Difeafes, is either a verminous Contagion, or noifome Exhalation contain'd in the Air, taking their Original from putrid Bodies, excited to Fermentation; and fo rarify'd and open'd by the ambient Air, whereby the virulent Atoms expire, and are drawn in by the Lungs; or elfe, being pent up for a fhort Time, break out, and fuffocate the vital Spirits, as flaming Brimftone does a living Infect."*

A little later on Browne continues with his thoughts on the route of the disease into the body, either through the stomach or pores of the skin. This latter mechanism of disease entry is key to a strand of medical thinking about plague for centuries; that atmospheric heat and human activity which raise body temperature cause the skin pores to open and so allow the malevolent pathogen access: *"The outward antecedent Matter arifeth from fome minute venomous Particles, or Effluvias of various Bodies, either received in at the Stomach, or immediately entering thro' the fmall Pores of the Skin where they perform fatal Effects, according to the Capacity of the Receiver; for it happens fometimes that a peftilential* Virus *is taken in, and by Virtue of a Robuft Conftitution forthwith excluded again; fo that what would prove detrimental to one Man's Health, little impairs another."*

Behind the warning by Vandernote [1569] to his readers in regard to '*refting, or waking, or mouing*' lies the notion of open skin pores through personal activity: *"When the peftilence ftrongly raigneth, then muft you beware of greate trauayle and labour, and fpecially in the open ayre. And when you will labour, fo fhall you doo* [sic] *it fafting and in a clofe ayre. And in time of Peftilence fhall you keepe you temporate in labour, for to* [sic] *much quietneffe, and to* [sic] *much labour is naught."*

And elsewhere: *"... and them that ferue in hot houfes: and all they that do lightly fweate thorowe fmall laboure or trauayle, or waxe hot, and all they that lightly waxe angrye, hote and fumous, al fuche are inclined unto the plage* [sic]*."*

On a slightly different tack, Thayre [1625] looks upon changes in atmospheric heat as problematic in terms of raising body temperature and subsequent impact on skin pores: *"When the temperature of the aire is changed from his naturall eftate to immoderate heate and moifture, then it corrupteth putrifieth, and ingendreth the Peftilence."*

Conversely, very cold weather, as opposed to the damp cold environments, could hold back the disease, and in the English translation of De Mertens' [1799] account of the 1771 Moscow plague he noted that when the really cold weather appeared then the disease did not spread, even to those handling bodies: *"The weather continued very cold until the middle of April, in confequence of which* the contagion became more fixed and inactive, attacking only thoſe who dwelt with the infeſted. In the peſt-houſe, the daily number of deaths did not exceed three or four..."

By the early part of the 19th century there had been a little more analytical thought going on (perhaps prompted by descriptions of the Moscow outbreak), so we find Pearson [1813] advising:

"*The activity of the contagion of the plague is promoted, weakened, or destroyed, by certain states of the atmosphere, and particularly by extremes of cold or heat.*
In the middle and northern parts of Europe, its malignity is subdued by a degree of cold at or below 32° of Fahrenheit; so that, in those countries, it never prevails epidemically *(though it sometimes exists* sporadically, *and in a mild form)* during the winter season. Its activity is in like manner said to be diminished, and sometimes destroyed, by an opposite temperature of the atmosphere, namely, by a high degree of heat, such as 90° and upwards of Fahrenheit's thermometer.*"

Another notion that existed, and which is exemplified by the following passage from Browne [1720], is that the individual could store up bad Humours (through poor diet, for example), and then heat from external sources or gained through physical activity could act upon these bad humours to bring on plague:
"*Within the Circumference of the Skin, an abſolute peſtilential Poiſon is ſometimes generated, which being incubated by the external Heat, and fermented by ſeveral outward Accidents, is at laſt maturated into this kind of virulent Matter that occaſions the Plague; for, undoubtedly, were a Man ſeparated from all Society, and lived in never ſo wholeſome an Air, yet ſuch a peſtilential Seed may ſpring up in the Body, capable of producing the ſame Effect, which an extraneous Matter occurring frequently doth; that is, receiving the Infection without Commerce, Contact, or Converſation with Perſons Infected.*"

In the following contemporaneous passage from Blackmore [1722] the connections and medical logic of the time are perhaps more obvious: "*... irregular Agitations and Conflicts, that ariſe in the Blood, either from a ſudden cloſing of the Pores of the Skin, or from luxurious Eating or Exceſs of Wine, or ſtrong Liquors, that by their crude or undigeſted Juices obſtruct the Strainers of the Body, interrupt the Circulation of the Blood, and hinder the Separation of the Humours requir'd for the Depuration of it, and preſerving it in a healthful State: Hence various Impurities are retain'd in the Maſs and prove detrimental to its Conſtitution, where continuing unexpell'd they contract an acrimonious, diſeaſy Quality, which agitates and diſturbs the Blood, and by that Means awakens and ſets at Liberty the Primitive innate Seeds before mention'd, by which Means Malignant Fevers are ſometimes produc'd.*"

When plague appeared in a victim then the onset of symptoms, and disease progression, could be extraordinarily rapid. De Chaulliac, who observed the Black Death in Avignon during 1348, was medical attendant of Pope Clement VI and one of the first European medics to seriously try and study the plague thought-fully. In his writings about the plague he concluded:

"It was of two kinds; the first lasted two months, with constant fever and blood-spitting, and of this people died in three days.
"The second lasted for the rest of the time. In this, together with constant fever, there were external carbuncles, or buboes, under the arm or in the groin, and the disease ran its course in five days. The contagion was so great that not only by remaining [with patients], *but even by looking* [at them] *people seemed to take it; so much so, that many died without any to serve them, and were buried without priests to pray over their graves."*

Textual descriptions of the onset of the disease, and thoughts on those particularly susceptible to catching plague, are constant over the centuries. Bullein in his *Dialogue* [1578] suggests: "*They which are smitten with this stroke or plague are not so open in the spirits as in other sicknesses are, but straite winded; they do swone and vomite yellowe cholour, swelled in the stomacke with muche paine, breaking foorth with stinking sweate; the extreme partes very cold, but the internall partes boiling with heate and burning; no rest; bloud distillyng from the nose, Vrine somwhat watrie and sometyme thick with stincke, sometyme of colour yellowe, sometyme blacke; scaldyng of the tongue; ordure most stinckyng; with red eyen, corrupted mouthe, with blacknesse, quicke pulse and deepe but weake, headache, altered voyce, losse of memorie, sometyme with ragyng in strong people. These and suche like are the manifeste signes howe the harte hath drawne the venome to it by attraction of the ayre, by the inspiration of the arters to the hart, and so confirming it to be the perilous feuer pestilentiall. This is most true, of this commeth foule bubos, antaxis and Carbuncles, Sores through putrifaction... Also this feuer is scant to bee recoured and almost past help when these Symptomatas do appeare...*"

In the following century Thomas Thayre [1625] tries to tease the symptoms out into a more ordered and structured format:

"The fignes that fignifie and declare a perfon to be infected with the peftilence.
The firft is, a great paine and heauineffe in the head.
The fecond is, hee feeleth great heate within his body, and the outward parts cold and ready to fhake, and is thirfty and dry therewithall.
The third figne, is he cannot draw his breath eafily, but with fome paine and difficulty.
The fourth fignes [sic] *is, he hath a great defire to fleepe, and can very hardly refraine from fleeping, but beware hee fleepe not. And fometimes watching doth vexe and trouble him as much, and cannot fleepe.*
The fift figne is, fwelling in the ftomacke with much paine, breaking forth with ftinking fweate.

The fixe figne is, diuers and heauy lookes of the eyes, feeing all things of one colour, as greene or yellow, and the eyes are changed in their colour.

The feauenth figne is, loffe of appetite, vnfauoury tafte, bitterneffe of the mouth fower and ftinking.

The eight figne is, wambling of the ftomacke, and a defire to vomite, and fomething vomiting humors bitter and of diuers colours.

The ninth figne is, the pulfe beateth fwift and deepe.

The tenth figne is, a heauinffe and dulneffe in all the body, and a faintneffe and a weakneffe of the limbes.

The eleuenth figne is, the vrine moft commonly is troubled, thicke like Beafts water and ftinking, but fmell it not if you loue your health; but oftentimes the water doth not fhew at all, efpecially in the beginning of the fickneffe, therefore truft not vnto the water, but looke vnto the other fignes heere aboue fet downe.

The twelfth and laft figne, and fureft of all other, is, there arifeth in the necke, vnder the arme, or in the flanke, a tumor or fwelling, or in fome other part of the body there appeareth any red, greenifh, or blackifh coloured fore, thefe are moft apparent fignes to the eye, that this perfon is infected with the Peftilence.

But take heed, be not deceiued: for oftentimes a perfon is ftrongly infected with the Peftilence, and hath neither Apoftume, Carbuncle, nor botch appearing in two or three dayes, by which time he is neere his death; therefore when a botch doth not appeare fpeedily, it is always an euill figne and dangerous."

De Mertens [1799] in his account of the 1771 Moscow plague gives a little more insight: "The fymptoms of the plague vary according to the different conftitutions of the perfons whom it attacks, and the feafon of the year in which it appears. Sometimes it wears the mafk of other difeafes, but in general it is ufhered in by head-ach, ftupor, refembling intoxication, fhiverings, depreffion of fpirits, and lofs of ftrength; thefe are followed by fome degree of fever, together with naufea and vomiting. The eyes become red, the countenance melancholy, and the tongue white and foul. In this ftate of things, the patients are fometimes capable of fitting up, and going about for fome hours, or even a day or two. They feel an itching or pain in thofe parts of the body where buboes and carbuncles are about to appear. During the height of the plague, many of the infected die on the fecond or third day, before thefe tumours have time to come out, and with no other external marks except petechiae or purple fpots, which appear a fhort time before death; in fome thefe fpots are altogether wanting. The buboes and carbuncles generally come but on the fecond or third day, feldom on the fourth."

In the early part of the 19th century, when plague is much less common, the American physician Daniel Whitney [1835] describes it this way: "The symptoms of this terrible malady very nearly resemble those of typhus. The great line of distinction between them is, that the plague is always attended with buboes in the groins and armpits, which are not to be observed in typhus. The attack, which is generally towards evening, commences with a feeling of great lassitude and weakness, coldness, but *soon succeeded by heat of skin, giddiness, with pain in the temples and eyebrows. The natural expression of countenance is changed into a wild furious look, or sometimes a look claiming pity, with a sunk eye and contracted feature. The pulse is small, hard and quick, and sometimes intermitting. Staggering, sudden and extreme prostration of strength come rapidly on, the speech faulters, the stomach rejects almost every thing, the tongue is white and moist; there is a constant inclination to void urine, the evacuations are highly offensive, and the patient is perfectly indifferent about recovery. Generally on the third day the buboes commence forming by excruciating pains in the groins and armpits, and the patient then frequently goes off delirious; or if he survives till the fith [sic] day when the febrile symptoms abate, he is almost sure to die from debility. But if the patient lives over the fifth day, and the bubo is fully formed, he is then considered as nearly out of danger."*

Reflecting Whitney's comment that the attack is 'generally towards evening' John Jones [1566] had this to say some three hundred years earlier: "*Furthermore the pulſe in this ague, is buſier in the night, then* [sic] *in the day, for the feuer is greater, and the pacient is ſhorter winded, and bretheth painfullier, and is verye thirſtye, for the pipe of the lunges and mouth ben drye, the toung is white or yelowiſhe, in the ouerpart, and blacke in the top and ſwollen, the ſpeach loſt, and taſt taken away...*"

Earlier mention was made of how ill Humours stored within the body were believed to make an individual more susceptible to a plague attack, but another widely held belief was that 'fear' itself could make someone more personally disposed to catching the plague. Joseph Browne [1720] intimates this when he says: "*... when a timorous Perſon hears a Relation of a malignant Small Pox or Plague, and the terrible Symptoms attending them, he immediately forms to himſelf a perfect Idea of that Diſeaſe, which rouzes in his Blood a malignant Ferment or contagious Seed, which ſeizes on the vital Spirit, and there the Conflict is begun and carried on, till one or the other conquers.*"

Perhaps a more telling passage comes from Bradwell [1636] about a hundred years previous, the work published in the same year as a serious London plague outbreak. What is fascinating here, however, is the medical logic – that a person's inner concentration on peril or fear leaves the external skin exposed to attack by disease: "*... of all the Paſſions, Feare is the moſt peſtilenly pernicious: And this it is: Feare enforces the vitall Spirits to retire inward to the heart: By which retyring they leave the outward parts infirme, as appears plainly by the paleneſſe and trembling of one in great feare. So that the walls being forſaken (which are continually beſieged by the outward ayre) in comes the enemy boldly; the beſt ſpirits that ſhould* [have] *expelled them having cowardly founded retreat; In which with-drawing, they draw in with them ſuch evill vapours as hang about the outward pores...*"

From around the same time the English version of Paré's work, circulating from 1649 onwards, similarly places importance upon the heart when under attack by plague: *"This peſtiferous poiſon principally aſſail's the vital ſpirit, the ſtore-houſ and orginal whereof is the heart, ſo that if the vital ſpirit proov ſtronger, it drive's it far from the heart; but if weaker, it being overcom and weakned by the hoſtile aſſault, flie's back into the fortreſs of the heart, by the like contagion infecting the heart, and ſo the whole bodie, beeing* [sic] *ſpread into it by the paſſages of the arteries."*

In *Plain and Easie Method* [1691] we have the following overview on how the body's 'Spirits' can be propped up through diet, wine, and personal confidence to overcome fear to repel the plague: *"Now becauſe the Spirits are commonly the firſt that receive Infection; We muſt fortifie them, that they may not eaſily admit the approaches of their Enemy, which when they are in full vigour and expanſion, they will repel, and as it were keep off at a diſtance; Therefore Wine and* Confidence *are a good* Preſervative *againſt the* Plague: *But when the Spirits, through fear, or want of ſupply, do recede, and are forc'd to give back, the Enemy enters, and firſt ſeizeth them, and thence gets into the Blood and Humors; Therefore much* Faſting *and* Emptineſs *are bad: But every one ſhould Eat and Drink at convenient Hours, in ſuch manner and meaſure, as may always keep the Spirits lively and chearful, and endeavour to compoſe his Mind and Affection againſt fear and ſadneſs."*

In his *Loimographia* [1666] the London apothecary William Boghurst was a clear advocate of good wine as both plague preventative and cure; regarding it as a cordial in its own right or, as in the recipe below, adding further ingredients such as borage and wood sorrel:

"I cannot diſlike the uſe of any good wine, as Canary or Rhenish, given only to fit persons in due tyme and quantity either for preſervation or cure, if it bee not juſt in the beginning of the diſeaſe, nor the patient very ſleepy or have a high feaver, or in hot weather, or bee a cholerick perſon, or frantick or light-headed, or have a great pain in the head, or thirſtineſs, or inflammations, or bee a child, in ſuch caſes it is to bee forborne; but to others free to bee uſed eſpecially for preſervation, for there is ſcarce a better Cordiall to bee found that growes in ſoe great a quantity amongſt all Simples, and doth much good in a plague tyme by expelling feare and melancholy, or you may mix a Cordiall with it as thus:

Rx. Vini Canarini lb. s.s. aqua boraginis, Cinnamoni hordeati, ana ℥ iiii Syr. e Succo Lujulae, berberorum, Idei rub., e Succo citri ana ℥ i aq. rosar. dam. ℥ ii M."

It is also interesting to look at other aspects of medical logic and reasoning that partly explain the treatments suggested. For example, in Bullein's *Dialogue* [1578] we have the following response from one of the characters concerning patients where plague sores and swellings do not appear: *"But often tymes the Plague sore will not appere; the very cause is this: Nature is to* [sic] *weake, and the poyson of the infection to* [sic] *strong that it can not be expelled, and this is moste perilous of all, when such a cruell conquerour doth raigne within the harte, the principall part of life, nowe possessed with death. The causes of this I haue declared before, with signes to the same; notwithstanding, consider two thinges: first, whether it is in bodies Sanguine and Cholerike, or theim whiche are Flegmatike or Melancholie, or not. The firste twoo, bloud is the cause, the seconde twoo aboundaunce of euill humours. Therefore let blood, where as it hath the victorie, and purge wheras other humours hath predomination or chief rule: in some men that haue verie stronge bodies, firste purge, then let bloud. Note this: that what side be infected let blood on that side; if it be aboue the hedde, open* Cephalica; *if it be vnder the armes,* Basilica, *or harte veine; if it be aboute the throte, then open* Melleola; *about the flankes, bealie, legges, &c., open* lecoriaria. *If thei are verie weake or yong, then boxyng is good to the necke, shoulders, backe and thighes; if the stomacke be full, then with speed vomet, and these thinges drawe the venome from the hearto and remoue the poison."*

About a hundred years later Willis' *London Practice of Physic* [1685] reasons that plague cases fall into two groupings, dangerous and less dangerous; the latter occurring where the victim is courageous and has 'Fortitude of Mind', which has resonance with the previous examples of fear making some individuals more susceptible to sickness: "*The Cafe is full of Danger if the Contagion fuddenly paffes into an univerfal Sicknefs, and makes violent Invafions; if an Haemorrrhagy* [sic], *or only a fmall Pain happen in the beginning of the Difeafe; if the Urine be thick, and troubled, the Pulfe unequal and weak; if a Convulfion or Frenzy presently follow; if the Vomitings, or Stools, are livid, black, or very ftinking; if the Pufhes at firft red, afterward turn black and blue; if the Carbuncles are numerous, if the Buboes firft arifing, difappear; if the Strength be caft down on a fudden; if the Countenance looks difmal, or turns black and blue; if with a cold Stiffnefs of the extream Parts there be a burning of the Vifcera, especially if thefe or moft of them happen in a Body very cacochymical, or in an unwholfome Seafon. On the contrary, the Difeafed are bid to be of good Courage, if the ftate of the Peftilence be lighter, and lefs dangerous; if the Difeafe happens in a found and robuft Body with a Fortitude of Mind; if Remedies are feafonably adminiftred before the Difeafe has feifed the whole Mafs of the Blood; alfo, if the Courfe of the Difeafe goes on with a conftancy of the Strength, a Vigour and Evennefs of the Pulfe, a Suppuration of Buboes, and a large Discharge of Pus, and with the abfence of horrible Symptoms: mean while, tho we may hope here all good, yet it is not free for us to be fecure, becaufe fometimes, with a laudable Appearance of Signs, Ambufhes are privily laid for Life; and, as from a reconciled Enemy, we fuffer moft feverely, when we feem'd to have efcap'd his raging Threats."*

One of the beliefs which many practitioners held was that the disease needed to be 'moved', as it were, from the heart and internal parts of the body to the exterior. Blackmore, in his *Discourse on Plague* [1722], gives us a sense of this where he speaks of the body's struggle to subdue the disease and then the appearance of buboes being absolutely crucial to the outcome of the patient, since this was the body's way of expelling the disease to the outward parts: "*The Reaſon why the firſt Inſults of a Plague of the violent and reſiſtleſs Kind are ſo fatal, is this, that in malignant Diſtempers there must be allowed the Space of ſome Days, till the Spirits and the vital Principles of the Blood have laboured, ſubdued, and digeſted the crude Matter of the Diſeaſe, and changed its noxious Quality, and brought it down to a weaker, and leſs offenſive State, by which it is prepared, and made fit for Expulſion; but ſuch is the Strength and Fury of this higheſt Kind of Contagion, that it gives Nature no Time to perform this neceſſary Office for its Safety; for immediately, in the Space of one Day, and frequently of two or three, it diſſolves intirely the Union, and Connection of the Blood, and turns it into an incoherent Puddle, by a general Putrefaction. This universal Mortification appears in blue and blackiſh Spots on the Surface of the Body, but deeper than the Skin... If any in theſe Circumſtances recover, it is when Nature, unable to digeſt the malignant Impurities, makes a hard Shift to expel them crude and unconcocted as they are, and casts them out into Malignant Swellings, on the Surface of the Body: But if it does not attempt this Relief, the Agonies of Death are ſoon to be expected, and in the mean Time, the Phyſician can only preſcribe ſuch cordial and reviving Remedies, as may, for a Time, cheriſh and comfort the Patient, whom he cannot cure.*"

There is a similar thread in William Northcote's *Marine Practice of Physic & Surgery* [1770] later in the century: "*The peſtilential poiſon diſturbs all the functions of the body; for unleſs it be expelled to the external parts, it is certainly fatal. Nor is this to be done as in other fevers, by large ſweats, ſtools, a flux of urine, cuſtomary evacuations of blood, or by bleeding at the noſe, either natural or artificial, for they rather haſten deſtruction. The ſalutary and critical excretion which perfectly ſolves the peſtilential diſeaſe, is tumors in the ſurface of the body, between the third and fourth day. That there is a poiſon contained in theſe tumours appears from hence, that if a ſurgeon opens any of them with his lancet, and then bleeds a ſound perſon therewith, he will be immediately ſeized with the Plague.*"

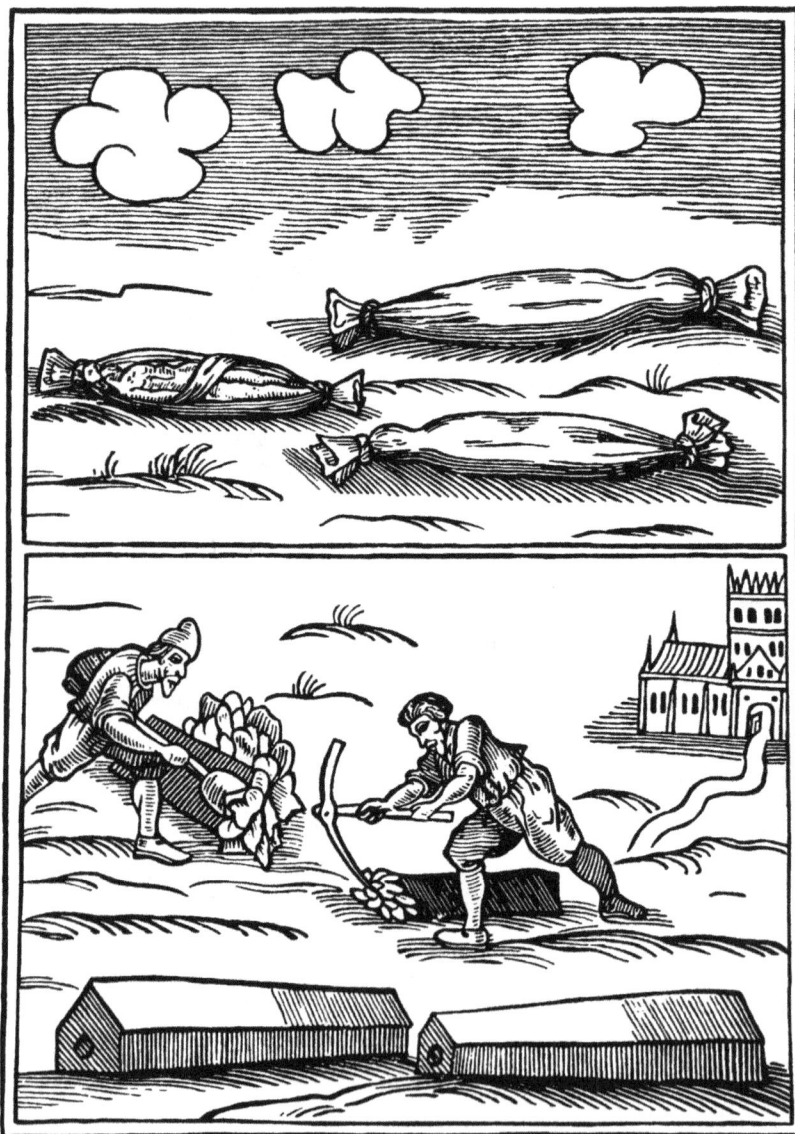

London had a number of serious plague outbreaks during the 17th century, in 1603, 1625, 1636 and then the Great Plague of 1665-6. An estimated 35,000 people died in the 1625 outbreak, but that nearly doubled in the 1665 epidemic and possibly reached 100,000 deaths, with 7,000 Londoners dying in a single week at the peak of the outbreak.

TREATING PLAGUE

Some insight into the way that physicians understood the disease, and also the functioning of the human body generally, helps us understand the reasoning behind many of the treatments pursued when trying to tackle a plague outbreak.

Not making some intervention to control the disease was most certainly regarded as erroneous, simply because plague was seen as too serious a disease to allow the body to try and mend itself. You can see this thinking in the Willis work, *London Practice of Physic* [1685]: *"In the Cures of moſt Diſeaſes, the chiefeſt Work is committed to Nature, to whoſe Failure Phyſick gives a helping hand: and the Office and Science of a Phyſician chiefly conſiſts in this, To wait fit Occaſions of giving Aid to her, when ſhe is at a Fault. But the Plague has this peculiar, that the Cure of it is not to be beſt* [left] *to Nature, but we muſt fight againſt it always with Remedies taken from Art; nor muſt we be here follicitous of a more feaſonable, and as it were, a milder Time: but we muſt get Medicines affoon* [sic] *as may be, and inſiſt on them at all Hours, and almoſt Minutes. But, becauſe when a Plague reigns there is need of no leſs care for driving away the Contagion, than that the Contagion receiv'd, be cured: therefore a Phyſician has a double Task; to wit, both that he take care for the Prevention of this Diſeaſe, and for its Cure."*

One of the best snapshots of medical thinking on the subject comes from a rather unusual source, John Howard's *Account of the Principal Lazarettos in Europe* [1791]. Howard travelled around much of Mediterranean Europe at the time, ostensibly to research the design and use of lazarettos which, perhaps not unsurprisingly, were often used during plague outbreaks to isolate the infected victims or suspected plague carriers from the healthy population. In normal times lazarettos were used as part of the general day-to-day quarantining of ships transporting passengers and goods. Just before departing for his trip from England Howard was asked by two doctors to canvas the opinions of Continental physicians about plague which he dutifully did, and devoted a section of his lazaretto book to the results.

The text extract which follows are responses to the series of questions he put to the Continental physicians: *What is the Method of Treatment in the firſt ſtage — what in the more advanced periods — what is known concerning Bark, Snakeroot, Wine, Opium, pure Air, the application of cold Water ?* In each case the name of the physician is followed by the location where they worked.

"RAYMOND [MARSEILLES]. *The diſeaſe is treated as inflammatory. No ſpecific has been diſcovered for it.*
DEMOLLINS [MARSEILLES]. *At the beginning, bleeding, vomiting, purgatives, diluents, refrigerants and antiſeptics are uſed; afterwards, antiſeptics and cordials, relatively to the temperament and ſymptoms.*

GIOVANELLI [LEGHORN]. *The plague, caufing always a difpofition to inflammation, and putrefaction, it is always proper to bleed proportionally to the ftrength, and to ufe a cooling regimen, with the vegetable acids. The repeated ufe of emetics is alfo proper, both to cleanfe the firft paffages, and to difpofe the virus to pafs off by the fkin. In the progrefs, it is neceffary to favour the evacuation of the virus by that iffue which nature feems to point at. Thus, either antiphlogiftic purgatives are to be given, if nature points that way; or fuppurative plafters are to be applied to any tumours which may appear. Epifpaftics to the extremities are proper where nature wants roufing. The vitriolic acid in large dofes has been found very ferviceable in the plague with carbuncles, as was proved in the laft plague at Mofcow. When the inflammation is over, and marks of fuppuration appear, the bark with wine and other cordials is proper. The furgeon's affiftance is requifite in the treatment of boils and anthraxes, which laft are feldom cured without the actual cautery.*

THEY [MALTA]. *In the beginning of peftilential fevers, bleeding is fometimes proper, and vomits almoft always. In their progrefs, frequent fubacid and cold drinks, the bark given liberally, and vitriolic acid, have been found powerful remedies when there was a diffolution of the blood.*

MORANDI [VENICE]. *In the firft period, evacuations according to the peculiar circumftances of the cafe are proper. In the fecond, bark mixed with wine; and opium as a temporary fedative. Pure air is very neceffary; and fire, as a corrective, with the burning of antifeptic and aromatic fubftances.*

VERDONI [TRIESTE]. *As foon as a Chriftian finds he has got the plague, he eats caviare, garlic and pork; drinks brandy, vinegar and the like, to raife the buboes. Upon thefe he applies greafy wool, caviare, honey of rofes, dried figs &c. to bring them to fuppuration.*

The Turks and Arabs drink bezoar in powder with milk, and other fudorifics, to expel the virus. They vomit, and poffibly a fecond time.

At Cairo they take opium, and cover themfelves with mattreffes in order to excite fweat; and though parched with heat and thirft, they drink nothing. They open the immature buboes with a red-hot iron.

At Conftantinople and Smyrna they eat nothing, and drink much water and lemonade. The Jews drink a decoction of citron feeds, lemon or Seville orange peel, and their own urine. They abftain fcrupuloufly from animal food.

In 1700 a phyfician in Smyrna found bleeding very ufeful. Another, in another year, cured the plague by bleeding and an antiphlogiftic regimen.

My brother in Cairo treated it like a pituitous biliary fever, with vomits, faponaceous attenuants, and antiphlogiftics, and fuccefsfully.

Some failors in Conftantinople in the phrenfy of the plague, have thrown themfelves into the fea; and it is faid that on being taken out, they have recovered.

My opinion upon the whole is, that the treatment ought to be relative to the particular conftitution of the year, and of the patient, by which the nature of the difeafe itfelf is greatly varied."

Two interesting points in the extract above: first, the mention from two of the physicians that vitriolic (sulphuric) acid was in use which really suggests that chemical pharmacy is perhaps in the ascendant and, second, Verdoni's mention of sea water which has a resonance with Tully's activities with sea water when stationed on Corfu (see page 56). However, a large proportion of anti-plague remedies and treatments remained in the vegetable domain.

Although the collective wisdom of the passage above indicates a cross-section of treatment approaches, a very generalized overview of the way in which plague was tackled is that patients were given a vomit or purge, or both, and that occasionally a pre-treatment draught of some medicine was given to 'prepare' the body for the purge as Barrough [1590] describes: "*A ſyrupe is good and profitable before a purgation, that they may be the eaſier and better obey the purgation. Therefore they are vſed of many Phiſitions to be giuen before medicines, although we haue not heard that old practiſers did obſerue it...*" A purge was often followed by a sweat (though treatment sometimes started with this), induced by a sudorific or alexipharmic. Even today we sometimes try and break an attack of flu by trying to sweat it out, so the principle is not unfamiliar. When the 'sweat' potion had run its course the patient was then given a regime of cordial medicines to help restore the body system along with some simple dietary items.

Despite the increasing use of chemical substances in remedies there were a large number of vegetable ones that could be employed, and a group of these regularly feature in old medical tracts on treating plague. In a 1615 edition of Dalechamps' *Histoire Generale des Plantes* there is the following outline list of plant remedies *Contre les fieures peſtilententielles* that includes oranges, scorzonera, violet, butterbur, barberry and cornflower (*aubifoin*) among others:

"*Eau d'Oranges beuë au poids de ſix onces.*
Violette d'Automne eſt ſinguliere contre les maladies peſtilentielles.
Decoction de la Scorzonera preparee auec le bouillon d'vn poulet guerit le fieures peſtilentielles, & eſt bonne contre les exhantemes qu'elle produit.
Eau diſtillee de Lentilles d'eau, prinſe en breuuage.
Racine du Petaſites ſechee & pulueriſee, prinſe en vin.
Fleurs de ſoucy ſeches, prinſes en breuuage.
Deux dragmes de la poudre de la Veronique ſechee, auec vne dragme de Theriaque prins en vin, & faire ſuer promptement le malade.
Syrop fait du ſuc aigre de Citron, beu.
Vne dragme de la poudre de la racine d'Angelique prinſe auec du petit vin, ou (ſi la fieure eſt trop ardente) auec eau de Chardon benit, ou Tormentille; & vn filet de vinaigre, & la donne-on ſeule, ou auec de la Theriaque.
Suc de Pimpinelle prins en breuuage.
Eſpine-vinette meſlé auec Iulep violat, prins en breuuage.
Eau des tendres fueilles du Cheſne diſtillees par vn alēbic de verre dans vn bain d'eau tiede, beuë.
Aubifoin prins en ſyrop de Chicoree."

While some of these plants and others might end up in a vomit or purge remedy it is interesting to look at the thinking behind the administering of purges and sweats and their actions on the patient's body. The following passage, from *London Practice of Physic* [1685], offers a good insight into the belief that the poisons of plague needed to moved from the region of the heart outwards: "*Medicines, in order to the Cure of the Plague, are either Evacuatives or Alexipharmicks. The Intention of the firſt is, that the Serous Latex in the Blood, and the excrementious Humours, which abound in the Viſcera, be ſent forth; and together with them, a great many Particles of the venemous Miaſm every where diſpers'd in the Body. Now theſe things are perform'd by Vomitories and Purges, whoſe uſe is rare, and only in the beginning of the Diſeaſe; alſo by Diaphoreticks, which at all times, as long as the Strength is able to bear them, are indicated in the Plague: for theſe evacuate more fully, and withal from the whole Body, and alſo by exagitating the Blood, they free it from Congelation, and in regard they move from the Center to the Circumference, they drive the venemous Ferments, and alſo the Corruptions of the Blood and Humours far from the Heart, and repell the Enemy from the Fort: but Vomitories and Purges evacuate leſs generally, and often by concentrating the malignant Matter, draw it inward, and fix it in the Viſcera.*"

We can partly follow the purging methodology through in the following passage from Thomas Thayre's *Treatise of the Plague* [1625]. He begins by referring to the widely held notion of the body abounding in superfluous humours which makes it prone to infection, but where the body is:

"*... free from ſuch ſuperfluous humors, there the infectious aire hath not ſuch matter to worke upon: And again, nature is more ſtrong and forcible to refiſt and expel a corrupt and infectious aire although receiued.*

Here the reaſon is apparant why one perſon is infected and not another. And very needfull it is eſpecially in this time of ſickneſſe, that this euill diſpoſition of the bodie be taken away and amended, by purging and euacuating of the peccant humors. For which purpoſe I will ſet downe a very excellent and approued potion, which purgeth the blood and disburdeneth the body of ſuperfluous humors both choler, flegme, and melancholie, opening attracting and evacuating the corrupt and vitious humors of the body, to the great comfort, help & eaſe of thoſe that vſe it with diſcretion, as I ſhall direct them: the making or compoſition whereof I haue here ſet downe.

But firſt taking this ſirrup three mornings before you purge, two ſpoonefulls euery morning, faſting after it two or three houres, and vſe your accuſtomed dyet as before."

[ED - To prepare the body.]
"*R. Oximell two ounces, ſir. de quinque radicibus two ounces, miſce.*"

[ED - The purging medicine.]
"*R. Good Rubarb two drammes, ſpicknard ſix graines, Sene halfe an ounce, Fenill ſeede, and Anniſſeede of each halfe a dram, flowers of Borage and Bugloſſe, of each*

halfe a little handfull; Water and Endiue and Fumitarie of each of them fiue ounces, and fo make your infufion.

Let this infufion be made in fome earthen ftuepot clofe couered and pafted that no breath or vapor goe forth, and let it ftand feuen or eight houres vpon fome imbers, or fmall coales, and but warme: after which time ftraine it forth and put thereunto of Diacatholicon one ounce, Diaphenicon halfe an ounce, Electuarium Succo rofarum halfe an ounce, *mixe thefe with the infufion aboue written, & this will bee a fufficient quantitie for three dayes, taking the third part the firft day, and on the fecond day the halfe of that which was left, and the other part the third day: take it early in the morning, and fleepe not after the taking of it, neither eate, nor drinke vntill it hath wrought his effect, and then take fome broath made with a chicken or a capon, and for want thereof with veale or yong mutton, as you can be prouided, with reafins* [sic] *of the Sunne ftoned, two or three dates, a little parfley put there-unto, and thickned with fome crummes of bread. When your potion hath done working you may take of this broath, and alfo a little of your meate fparingly, and in the euening make a light fupper with a chicken, or a rabbet, or fuch like meat that is light & eafie of digeftion, yeelding good nutrimēt: the next day early, take another part of your drinke, and vfe your felfe as the day before. And like wife the third day, take that part of your potion that remained, and vfe your felfe as before taught."*

Sherwood [1641] recommends vomits for those who can take them but for those in a weaker state suggests less potent remedies and, failing that, blood-letting, enemas (*clyster*), or sweats:

"Whofoever fhal perceive their bodies infected with the Plague, let them take on the firft day of the ficknefle the vomit... And after it hath done working with them, they fhal find themfelves as well as ever they were in their lives: for it clenfeth the ftomack and bowels from al corrupt humours, which is one of the chiefeft caufes of the ficknes. But if the fick be weak and cannot bear a vomit, it fhall be good to give him one dram of the forefaid pillulae peftilentiales, *or inftead thereof one dram of Aloes, you may give it either in pill or in potion, according as the fick can beft take it, and in the workking* [sic] *of it let him drink fome warm broth.*

But if it be fo, that this courfe hath been neglected the firft day, or beyond the time of 24 houres, it will bee in vaine to ufe it the fecond day: Yea, it will bee, dangerous, feeing that the infection is difperfed by the bloud throughout all the Veines of the bodie. Therefore on the fecond day of Vifitation it fhall bee good to draw from the Median Veine of the arme fo much bloud as the patient can endure to bleed: and if the ficke hath not gone to the ftoole during the time of his ficknefle, you fhall give him either before or after bleeding this Clyfter. Take of Beets, Violet leaves, Burrage, Bugloffe, Scabios, of each one handfull, French barley one ounce; boyle all thefe in a fufficient quantitie of water untill it be halfe confumed, then ftrain it, and take three quarters of a pint of the decoction, and put to it the Electuary of Hierapicra five or fixe drams, oile of Rue one dram, red Sugar one ounce, the yolk of an egg, and a little falt; fo make you a Clyfter thereof, and adminifter it bloud warme. Alfo you may adminifter to the fick this Clyfter. Boyle an handfull of Rue, in a pint of poffet drinke, and put to it a piece of fweet butter, a little honey, the yolke of an egge, and a thimble full of falt; make a clifter and adminifter it bloud warme.

But if that the fick amend not upon this courfe taken the fecond day, or that this means hath not been ufed, but that hee continueth fick untill [sic] the third and fourth day, fo that the infection hath taken hold of the vitall fpirits, Then keeping him warme in his bed, you fhall ufe this Cordiall to fweat with all. Take of the water of Scabios, Burrage, Bugloffe, and Angelica, of each halfe an ounce, the Electuary of egges two fcruples, or one dram, of Bole Armoniak one fcruple, Syrrup [sic] of Rofes halfe an ounce, make it into a potion, and let the fick drink it up at once or twice: two or three fpoonfuls hereof is fufficient for a child. Or the poorer fort may take two peny-worth of Treacle or Mithridate, in a quarter of a pint of Dragon water. With either of thefe Medicines you may fweat the fick, untill [sic] fome tumour doth appeare, or that he commeth to know himfelfe amended. For this is the laft medicinall [sic] refuge we have in the cure of the Plague. If you can, keep the fick from drinking and fleeping for the fpace of three houres untill the medicine hath done workking [sic]. But if you cannot, let the patient drinke a little Limon poffet, made with fome Marigold flowers, and Harts horn.

And if fignes of amendment doe appeare, doe not take him out of his bed, or let him coole fuddenly: But let him fweat on gently of his own accord, for it is natures fweat following the medicine, which will doe him more good than a Kingdome."

A little later Sherwood says: "If any that are ancient or weak fhall be infected with the Peftilence, it fhall not be neceffary to give them any purge, vomit, or fweat, or to let them bloud; becaufe they cannot beare the loffe of fo many fpirits as are fpent by fuch evacuations. Therefore you may lay upon the pit of the ftomack of the ficke a young live puppy, and if the fick can but fleep the fpace of three or foure houres, they fhall recover prefently, and the dog fhall die of the Plague. This I have known approved; and I do believe that it will be a cure for all leane, fpare, and weake bodies both yong [sic] and old: provided, that the dog be yonger [sic] then the fick."

If a plague victim was unlucky enough to be struck down while supping then immediate administering of a vomit appears to have been recommended, at least by Barrough [1624]: "And if the Peftilence do inuade any man at his dinner time or fupper time, when the ftomach is filled with meate, then he muft vomite ftraightway. At the laft, when the bodie and ftomach is emptied, you muft by and by minifter fome medicine that can refift poifon, that it may draw the poifon to it, and call it backe from the heart: for that is the propertie of fuch medicines."

In his short pamphlet on advice for the poor during the plague Thomas Cocke [1665] recommends: "For all Perfons fo foon as they find themfelves ill or infected... is imediately [sic] to Vomit... The beft way of Vomiting will be to put two or three Ounces of Oximel of Squill, into a Pint or Quart of Carduus Poffet, and drink it all off, This will coft about *fix pence*."

After the vomit Cocke recommends: *"Within one quarter of an hour after, whether you Vomit or no, go into a warm Bed and fweat Fifteen or Twenty hours (if it can poffibly be endured) with two or three pennyworth of* London Treacle, *(more or lefs, accordig* [sic] *to to* [sic] *the Age and Conftitution of the party) diffolved in five or fix Spoonfull of warme Vinegar; and as often as they Thirft during their Sweat, let them Drink freely hot Poffet-Drink, or Mace Ale, a little Rofemary and Sage boiled in it, and Drink no other Drink for two or three days; nor any cold Drink for four or five days or more. An hour or two after the Sweat is over, and the Body well dried with warm Cloaths, they muft wafh their Mouths and Hands with warm Water and Vinegar, and then (if their ftomach will permit) they may refrefh themfelves with fome convenient food; as Mutton Broth, Egg Caudle, Water-Gruell, or Panada, with a fprig of Rofemary, Mints, or Thyme, boiled in it."*

Again, in the following Willis [1685] extract, one sees the notion of making the body push the disease outward, while at the same time outlining the sequence of treatment steps: *"If the Plague happens in a Body that is not well purg'd, and prone to Vomit, prefently let a Vomitory be taken; the Operation of which being ended, prefently Diaphoreticks being giv'n, let a Sweat be procured, and let the fame be continued as the Strength will bear; afterward let it be often repeated: moreover, Alexipharmicks muft be ufed almoft every moment, till the Venom be wholly fent forth by the eruption of Pufhes, Carbuncles, or Buboes; neverthelefs, in the mean while, let refpective and proper Remedies be oppos'd to the Symptoms chiefly preffing; but efpecially for the cure of Buboes and Carbuncles, let fit Aids be taken from Chirurgery. The whole ftrefs of the Bufinefs relyes on thefe two chief Intentions, that the peftiferous Venom be by all means expelled from within, outward; and then that the Return of it, being expulfed, be with an equal Diligence prevented."*

Willis outlines the use of alexipharmics – medicines used to neutralize a poison but raising a sweat – in cases of plague as follows: *"As to Alexipharmicks, which are faid to refift the Venom of this Difeafe without a fenfible Evacuation, they are for the moft part fuch whofe Particles are neither very agreeing with Nature, that they turn to an Aliment, nor are fo contrary to it, that they ftimulate an Excretion. The fame being inwardly taken, and refracted to moft minute Parts, they, by their Corpufcles, infpire as it were with a new Ferment, the Blood and Juices flowing in the Veffels and Vifcera, and by moving them gently, and keeping them in an even mixture, they free them from Coagulation and Putrefaction: by the fame gentle Agitation they diffipate from each other, and hinder from maturation the Particles of the Venom beginning to be gathered thick together. Laftly, by praeoccupying* [sic] *the Blood and Spirits, they defend them from the Ingreffions of the peftilent Character: of thefe Remedies fome Simples are recommended, as* Rue, Scordium, *&c. but thofe are efteem'd far better, which are more compounded; wherefore* Treacle, Mithridate, and Diascordium, *whereof some consist of at least fifty Simples, are accounted Medicines so compleat in all Respects, that it is esteem'd a Crime to omit, in the making of them, ev'n one Plant; the Reason haply is, That many things being put together make a Mass, whose diversifyed Particles, being exalted by a long Digestion, cause a greater Fermentation in our Blood and Humours."*

As mentioned towards the end of last extract a remedial potion could contain a large number of individual ingredients. In regard to alexipharmic-related sudorific remedies James [1747] outlines some of the best vegetable sudorific ingredients as follows: "Sudorifics, *by whose Operation a sensible Moisture is perspired through the Cutaneous glands... Of the Vegetable Kind the most efficacious for this Purpose, are the Roots of a very acrid, penetrating, oily Taste, as those of Angelica, the different Species of Master-wort, Butter Burr, Elecampane, Lovage, Swallow-wort, Valerian, Contrayerva, Virginia Snake-root, Woods of Guaiacum and Sassafras."*

Elsewhere James advises that there is little point in giving sudorifics, particularly if they were very active ones, unless: "*... the porous Substance of the Skin be sufficiently open and lax, or unless the Blood be enough diluted. Wherefore if any one, in the Cure of a Disease, thinks Sweating required, it will be necessary for him to give the above-mentioned Sudorifics with a sufficient Quantity of some Liquid to dilute the Blood, for Example, a weak Tea, or a Decoction of Barley...*"

Alexipharmics were much more potent than simple diaphoretic, perspiration-increasing, remedies and were not always approved of as James suggests here: "*Alexipharmics... which frequently do considerable Mischief; and indeed there is Nothing in which the lower Class of Practitioners in Physic make more Errors, than in the Use of Alexipharmics, which I have frequentlly known exhibited to young People, of plethoric Habits, in the very Beginning of Fevers, and even without previous Evacuations.*" Quite what the reference 'lower class of practitioners' relates to this writer is uncertain, but perhaps it is aimed at the apothecary trade.

Evacuation of the body by vomits were not always used or appropriate as Richard Mead points out in the following extract from his *Discourse on the Plague* [1744] and also comments on treating plague sores at the bottom:

"*When the Fever is very acute, a cool* Regimen, *commonly so beneficial in the* Small-Pox, *is here still more necessary. But whenever the Pulse is languid, moderate Cordials must be used.*

The Disposition of the Stomach and Bowels to be inflamed, makes Vomiting *not so generally safe in the* Plague, *as in the* Small-Pox. *The most gentle* Emetics *ought to be used, none better than* Ipecacuanha; *and great Caution must be had, that the Stomach or Bowels are not inflamed, when they are administer'd: for if they are, nothing but certain Death can be expected from them: otherwise at the beginning they will always be useful. Therefore upon the first Illness of the Patient it must carefully be considered, whether there appear any Symptoms of an Inflammation having seized thefe Parts: if there if there are any Marks of this, all* Vomits *muft be omitted: if not, the Stomach ought to be gently moved.*

The Eruptions, *whether* glandular Tumors, *or* Carbuncles, *muft not be left to the Course of Nature, as is done in the* Small-Pox; *but all Diligence muft be ufed, by external Applications, to bring them to* Suppurate. *Both thefe* Tumors *are to be treated in moft refpects alike. As foon as either of them appears, fix a* Cupping-Glafs *to it without* fcarifying; *and when that is removed, apply a* fuppurative Cataplafm, *or* Plafter *of warm Gums."*

One aspect not elaborated upon so far in plague treatment, although mentioned in several of the previous extracts, is the activity of blood-letting. When reading general medical literature through less enlightened centuries one is left with the distinct impression that no treatment of disease or a medical condition was considered complete without blood-letting. The patient might very well be on death's door but he may still have been bled, while Galenic doctrine, as Hancocke [1723] puts it, "... *advises to let Blood 'till the Patient faints.'*" Neither of which would help the patient one iota. In the case of plague treatment there was the added danger that *Yersinia*-infected lancets would be used without being sterilized, and the widespread use of blood-letting simply spreading the disease among otherwise healthy parts of the population.

Blood-letting appears to have been employed with different medical intentions. While its general use seems to have gone hand-in-hand with the notion of balancing the Humours and evacuating the patient of infected blood Robert James, in *Pharmacopoiea Universalis* [1747], gives another use, heat reduction: "*Hence, alfo, appears the Reafon, why Venefection is the moft infallible Method of diminifhing the Heat of the Body; becaufe, by leffening the Quantity of Blood, its Attrition in the Veffels, on which the Denfity of the Humours depends, is proportionally leffened."*

However, when you relate comments in previous sections of this book, about the understanding of types of fever, then James' following comment would make absolute sense to a 18th century medic in terms of blood-letting for plague victims: "*A Fever is an intenfe Commotion of the Blood, excited in order to remove and expel what threatens the Deftruction of the Body...*"

William Bullein was an advocate of blood-letting at the start of a plague attack, and in his *Dialogue* [1578] gives some interesting reasoning for this advice: "*Blood muft be letten in the beginnyng of the sickeneffe. For example, like as a pot if clenfed of the fcum or fome in the beginning when it plaieth on the fire, and thereby the liquor is cleanfed within the potte, euen fo blood lettyng and pilles doe helpe and cleanfe the Peftilence when it beginneth firfte to boile within the bodie. Howbeit, certaine people maie not bleede, af women whiche haue their times aboundauntlie, or menne hauing fluxe of the Hemoroides, children verie young, or people weake and aged."*

In the curation of plague Kellwaye [1593] says medical help should be sought early, stating that: *"There are foure intentions required for the curing thereof: that is by bloud letting, Cordials, fweat and purging..."*

More precise details of *'Who ought to bleede'* are given by Thayre [1625]: *"Firft, when any feeleth himfelfe ficke or euill at eafe, if the fickneffe begin hot with paine in his head, if he be of a fanguine or cholericke complexion, or hath a plethoricke body, that is, a body full of humors, large veines and full, let euery fuch perfon in any wife be let bloud in the liuer veine and right arme. And if there fhould be felt any forenes in any fide of the body more then [sic] the other, then let him bleede in that arme on the fide grieued, which being done; let the Chirurgion decently binde vp his arme; and if the perfon be weake, then let this be done in his bed, and with fpeede let him take one of the foure medicines before fet downe in this booke for the cure of the peftilence, the quantity and the manner is there fet downe. Let him receiue his medicine warm, and procure him to fweat, which if hee cannot eafily doe, then muft you fill fome bottels with hot water, and fet them in the bed about him, by which meanes you fhall caufe him to fweat fpeedily."*

Not all physicians were convinced that blood-letting was an appropriate course of action in the disease. The renowned 16th century physician and surgeon Ambroise Paré, who had made great strides in the treatment of gunshot and other wounds of war, was very much against blood-letting in the plague, arguing that it left the patient weaker. The 1649 English version of the 1575 French original says: *"In the year of our Lord God 1566, in which year there was greaat [sic] mortalitie throughout all France, by reafon of the peftilence and peftilent difeafes, I earneftly and diligently inquired of all the Phyficians and Chyrurgians of all the Citties (through which King* Charls [sic] *the Ninth paffed in hif progrefs unto* Bayon*) what fuccefs their patient had after they were let blood and purged; whereunto they all anfwered alike, that they had diligently obferved, that all that were infected with the Peftilence, and were* lett bleed *fom quantitie of blood, or had their bodies fomwhat ftrongly purged, thenceforwards waxed and weaker, and fo at length died; but others which were not* let blood *nor purged, but took cordial Antidotes inwardly, and applied them outwardly, for the moft part efcaped and recovered their health..."*

Apothecary William Boghurst [1666] too, was most decidedly against blood-letting and in a summation of treatments to cure the plague says: *"Therefore, firft, the total avoyding of bleeding, purging, and vomiting as most pernicious and deftructive by what meanes foever procured, or by whomfever practized [sic] or commended, by writing or otherwife; for though there bee 30 or 40 Authors which have commended bleeding, there are as many or more which condemn it, befides continuall experience."*

In *London Practice of Physic* [1685] we again find blood-letting cautioned against, while a range of alternative courses of action are suggested: *"The Aids which belong to Chirurgery are Bleeding, which ought to be uſed ſeldom, and with great Caution in this Diſeaſe, becauſe the Blood being too much exhauſted, and the Veſſels falling, a Sweat is not ſo eaſily procured; the place of this is better ſupplyed by Cupping-Glaſſes with a Scarification; for this, and Veſicatories are aptly uſed for drawing forth the Venom. Moreover, againſt Buboes, Carbuncles, and malignant Ulcers produc'd by them, Cataplaſms, Fomentations, Plaiſters, Liniments, and many other things, to be apply'd outwardly, are taken from Chirurgery, with which ſome poyſonous things, as Drawers of Venom, are preſcribed to be mix'd by ſome: wherefore, Preparations of Arſenick, viz. its Oyl, and Balſam, are recommended in this Caſe, as of excellent Uſe and Efficacy."*

If physicians could not agree among themselves in regard to blood-letting the same might be said for overall courses of treatment. While one might dive in straight away with a spot of blood-letting another would start with a purge or sweat, or vice versa, and incorporate blood-letting at another time. Kellwaye [1593], for example, didn't think purges should be given until the third or fourth day, yet most physicians would probably have expected many of their patients to be dead at that point. Remember, also, a point previously covered, that fevers were grouped together as a symptom to be treated, Blackmore [1722] saying that plague fever was of the same nature and kind as malignant fevers: *"... and differs from them only as it is more Fatal and Contagious, the Method and Medicines, which are moſt Effectual for the Cure of ſuch Fevers, will likewiſe be moſt uſeful for the Cure of the Peſtilence."*

By the Great Plague of London the Royal College of Physicians of London had decided that bleeding, purging and vomiting were not wise courses of medical treatment in plague, pronouncing in their pamphlet *Certain necessary directions as well for the cure of the plague* [1665] that: *"Theſe three great Remedies rarely have place in the Plague, but are generally dangerous, (and moſt of all, Purging by any ſtrong Medicines) and therefore not to be uſed but upon ſome extraordinary urgent indicant or juſt occaſion, and with the greateſt caution, which onely* [sic] *an able Phyſician can judge of..."* and they further expressed the opinion that: *"... the poiſon is beſt expelled by ſweating provoked by poſſet-ale and London treacle"* mixed, while the patient was to be: *"... put to bed to ſweat, well covered in a blanket, without his ſhirt, for twenty-four hours, every fifth hour renewing his cordial, but in half the quantity, between-whiles refreſhing him with poſſet-drink, oatmeal caudle, or thin broths, made jelly-wine, or hartſhorn jelly..."*

To a certain extent William Bullein had beaten the College of Physicians to a sweating regimen almost a hundred years earlier. In his *Dialogue* [1578] he outlines the general treatment that should be undertaken, beginning with a sweat: *"But in the peſtilence tyme, one beyng infected therewith, let hym ſweate by warme thinges, as hot tiles, &c.; and let not the pacient eate, fleepe, or drinke; and eate light meates, as Henne, Capon, Cheken, Partriche, eating often and little at once, with faufe made fharpe of veneger, Oringes, fharpe Limondes, or Sorel; and in the firft day of the fickenef, that the pacient bee kept from fleepe by talkyng, fprinklyng of fwete water, rubbyng of the bodie, as nofe, eares, or foft pullyng of the eares, as thei may be fuffered, or a Sponge dipped in Vineger applied to the nose; and if vehement drineffe or heate dooeth approache, then drinke the Syruppes lafte rehearfed, and haue the chamber cleane kepte, and alfo parfumed fower tymes of the daie. Beware of ftincke; let the perfumes be made with* Olibanum, Maftike, wood of Alooes [sic], Benjamin, Storax, Laudanum, Cloues, Iuniper, *or fuche like, and fprincle all the chamber about with vineger; rofes in the windowes, or greene branches of Sallowe or of Quinces are good, fprinkled with Rofe water and Vineger."* The mention above of 'hot tiles' is mirrored in the College of Physicians text which, instead, recommended the deployment of hot bricks.

A typical sweat remedy is the following one from Willis [1691]: *"To provoke fuch a Sweat: Take of* Venice Treacle, *One Dram; diffolve it in three Ounces of* Carduus Water; *add a Spoonful of* Syrup of Peftilential *Vinegar; Or take a Draught of Poffet-Drink, made with Peftilential Vinegar; In which boyl a few* Petafitis *Roots: To promote and continue the Sweat, take Poffet-Drink with* Meadow-Sweet, *or elfe with* Carduus, *or* Marigold-Flowers *boyl'd in it."*

Whether it was the posset or the ingredients in the pestilential vinegar which made the above recipe 'expensive', Willis follows up with a cheaper one for poorer people: *"Take* Celandine *and* Rue, *of each one handful,* Marigold Flowers *half a handful; boyl thefe in a Quart of* White-Wine *Vinegar; ftrain it out, and keep it in a Glafs Bottle; give two or three Spoonfuls; in which diffolve of* Venice Treacle, *or* Mithridate, *one Dram. This will provoke Sweat very powerfully."*

Another sudorific he gives for the poor was: *"Seeds of* Rue *pouder'd, and one Dram of it mix'd with half a Dram of* Treacle, *diffolv'd with* White-Wine, *is accounted an excellent Sudorifick."*

After the sweat Willis suggested administering the following, although this appears to contain some potentially expensive ingredients too: *"Take of* Harts-Horn *rafp'd, and* Ivory, *of each three Drams; a* Pearmain *fliced,* Woodforrel *half an handful: Boy'l thefe in three Pints of Water, till a third part is wafted; Strain it on two Ounces of* Conferve *of* Gilloflowers, *or* Woodforrel, *or* Red Rofes: *Let it infufe an hour: then ftir it, and ftrain it out: Give a Quarter of a Pint warm."*

Natural vomiting in the plague victim, as opposed to that induced by the physician's potion, had to be dealt with before a sweat could be administered, and again Willis advises: *"If the Party is much opprefs'd at his Stomach, and ftrains to vomit, or elfe with vomiting throws up bitter and ftinking Matter; let him prefently take a large draught of* Carduus, *or* Camomile *Poffet-Drink, and in it either half a Dram of Salt of* Vitriol, *or two Ounces of Liquor of* Squills, *and with his finger or a Feather, fetch up what is contain'd in his Stomach...*"

Having expelled whatever was in the stomach the patient could then be administered a sweat, Willis listing several as follows:

"*Take of Compound* Scordium Water *two Ounces;* Treacle-*Water half an Ounce,* Venice Treacle *one Dram,* Salt of Wormwood *one Scruple,* Spirit of Vitriol *Six Drops:* Mingle them.

Take of Butter Burr-Water *three Ounces,* Venice Treacle, Diafcordium, *of each one Dram, Peftilential* Vinegar *one Spoonful:* Mix them.

Take of Carduus Water Four Ounces, Peftilential *Vinegar* One Ounce, Peftilential Extract *One Scruple;* Mix them."

Nathaniel Hodges, a medical contemporary of Willis, identified that some patients failed to respond to administered sweating potions. The following extract is from Quincy's 1720 edition of Hodges' *Loimologia,* originally published in 1665 at the time of the Great Plague of London: "*Some indeed of a very dry Temperament, or from a Confumption of their Humidities by the febrile Heat, do not eafily get into a Sweat; fuch therefore I ordered liberally to drink of a medicated Poffet-Drink; in order by this Means both to render the vifcid Humours more fluid, and contemper and affwage the feverifh Heat.*

The Milk with which this Poffet-Drink was made, was turned with two Parts of Ale, and one Part of Vinegar, in which had been boiled the Roots of Scorzonera *and* Butter-bur; *the Leaves of* Baum, Scabius, *and* Wood-forrel; *the Flowers of* Borage *and* Marygolds; *the Rafpings of* Ivory *and* Hartfhorn, Carduus *and* Coriander *Seeds.*"

The sweats were maintained for two or three hours at a time and if the patient could face eating food he might be given: "*... Bread foaked in Wine, poached Eggs with* Juice of Citrons, Pomegranates, *or* Elder-Vinegar, *as alfo cordial Waters, Broths, Gellies, and fometimes alfo generous Wines.*

The Broths then ufed were made by boiling in Chicken-Broth the Roots of Scorzonera, *the Leaves of* Pimpinel, Meadowfweet, Wood-Sorrel, Borage, *Rafpings of* Hartfhorn, *and* Dactyls, *with a Piece of White Bread, and a little* Saffron *tied in a Nodule; and the Fat was not taken off unlefs in a Loofnefs or Loathing at Stomach...*"

Blackmore [1722] was of the opinion that sweating should not be used at first, and appears to veer away from fierce sweat-inducing potions:

"*... the judicious Phyfician will firft rather reftrain great and profufe Sweats, by aftringent, cooling, and diluting Remedies, than promote them by warm and generous*

Remedies; and for that Purpofe I have often us'd with great Advantage, the following Decoction,

Take of Root of Sorrel, Dandelion, Fennel, each an Ounce, of Candied Eringo Root, an Ounce, of Seeds of Purflain, fweet Fennel, each a Dram; boil, them in f. q. of Water, to a Quart; add to it, when ftrain'd, of Treacle-Water, three Ounces, of Syrup of Lemons, two Ounces; drink of it a Quarter of a Pint, after each Cordial Bolus, before prefcribed.

But ftill the Phyfician will take care, that Cordial, or Acid, and attempering Medicines, may be given together in that Proportion, which the Symptoms and Neceffities of Nature do require, that is, where there are great Faintings, and the Pulfe is low, labouring, and depreffed, Cordials are demanded in a greater Quantity, and the others in a lefs; and when the Pulfe have more Force and Regularity, then the cooling and aftringent Remedies fhould be us'd in a greater Proportion."

Elsewhere in the Blackmore work there is paragraph in a similar vein to that above which highlights one of the physician's chief skills, that of adjusting a course of treatment or remedies to the temperament of the individual patient. Indeed, in herbal recipes you will sometimes see the phrase '*mix according to art*' or something along similar lines. In the 21ˢᵗ century that phrase rather passes us by, but in former times making up medicines was regarded as a craft and art, and apothecaries and physicians were expected to know the qualities and potency of every individual ingredient that went into their medicines and, if necessary, know about weaker and stronger substitutes so that the medicine suited the temperament of the patient. Regarding cordial medicines, and by inference to all other medicines used for curing any illness, James says they should be: "... *augmented or diminifhed, or given more or lefs frequently, according to the various Circumftances of the Patient; always fhewing a juft Regard to this ftanding Rule, which fhould govern the Phyficians Conduct, that when by their Weaknefs and Depreffion the Pulfe appears to be below, or by their Height and ftronger Strokes and Vibration, to be above the Standard of Nature, that is, that equal, foft and fteady Manner of beating, which they are obferv'd to keep in a healthful State, then the Medicines ought to be different; in the firft Cafe, fuch fhould be directed as have a fufficient Force and Vertue to raife and enliven the pulfe; and in the laft, fuch as are proper and effectual to lower and reduce them to the Standard before-mentioned..."*

Returning to sweats and sudorifics, Hancocke [1723] recalls Diemerbroek's use of them in the previous century as follows:

"*Sudorificks are looked upon by fome of the beft Phyficians, as the fafteft, quickeft, and moft proper Cure for the* Plague.
I will begin with Diemerbroek, *who is fuppofed by many to have wrote the largeft, and beft of the* Plague... *He tells you what his Cuftom and Method was. In the firft, fecond, and third Day, he gave 'em Sudorificks; if they vomited 'em up, he repeated 'em. If they were hard to Sweat, he put more Covers upon them. If yet the*

Fever increaſed, he repeated his Sudorifick the ſecond or third Time, nay, even ſometimes to the fourth or fifth Time. This Method ſhews, what Opinion he had of Sweating in the Plague."

Where sudorifics did not work to remove the 'venemous infection' the next course of action according to Thayre [1625] was a purge, although he cautions against this if the patient had developed plague sores:

"A purging potion of great vertue, that expulſeth all venemous and corrupt humors from the body.
Here I warne all men that they meddle with no purging medicine when the botch or carbuncle apeareth, and groweth towards ripeneſſe; for ſo ſhall they drawe the venome in againe, which nature hath put forth before.
R. Leaues and flowers of holy Thiſtle, Scabious, Turmentil, three leaued graſſe, of each a little handfull, Gentian, Tamarims, of each two ſcruples, good Rubarbe one dram: water of Bugloſſe and endiue, of each one ounce and a halfe, Sene three drammes, water of Scabious one ounce, flowers of Borage a little handfull, make your infuſion, which being done, put thereto diacatholicon halfe an ounce, Manna halfe an ounce, ſir. Rof. ſolutiue one ounce.
This potion hath a moſt excellent property in purging the body from venemous and corrupt humors, as the learned may iudge at the ſight thereof. This potion muſt be taken of the patient the ſecond or third day at the furtheſt after his ſweating, when no botch appeareth.

From France, Thiebault [1545] gave the following purge when others failed:

"Autres purgations bien experimentees pour prendre quād on void quil ny a nul remede &c.
Item prenez deux onces de ius de ſurelle / & autant de verbena ou de plantain / & eaue roſe une once / camphre & bolus rouge / de chaſcun demie dragme: mis tout enſemble ſoit dōne au paciēt tiede / icelui bruuage & fort refrigeratif: & chaſſe la peſte incōtinent de lentour du cueur ..."

And then from fellow Frenchman Royet [1583] there is this vomit which includes horseradish and orache seed: "*Prenez de la decoction du refort, ou de ſa ſemence, & ſemence d'arroche, de chacune trois drachmes, demye once d'oxymel, & autant de ſyrop aceteux, & en donner à boire au malade en bonne quantité vn peu tiede.*"

Perhaps understandably anyone with a high fever, or vomiting fluids and undergoing sweating, would be dehydrating heavily and likely to develop a thirst, and even here the drinks administered had a remedial aspect to them rather than the patient simply being given plain water. Kellwaye [1593] provides recipes for two such refreshing juleps:

"Take Syrrop of Ribes, Sorrell, Nenuphare, of either one ounce. Iuice of limons, one ounce. Sorrell water, eight ounces.
Mixe all thefe together, and take two or three fponefuls thereof often times, which will both comforte the hart and quench thirft.

And if in the time of his fweate he be very thirftie, then may you giue him to drinke a Tyfane made with water, cleane Barly, and Lycoris fcrapt cleane and brufed, boyle them together, then ftraine it, and unto a quarte of the licquor ad three ounces of fyrop of Lymons, and giue thereof at any time, fmall beere or ale is alfo tollerable..."
It is interesting to reflect that lemons would have been a highly priced item for 1593, so these remedies would almost certainly have been for the wealthy.

Royet [1583] again, around the same period, outlines a number of thirst quenching options in a section entitled *'Du boire du malade peftiferé'*, and includes wine, oxymel, and a mixture of honey in water with cinnamon added: "*Si le malade a grād' fieure, & ardēte, il ne boira aucunemēt du vī, s'il ne luy furuiēt defaillāce de coeur: mais en lieu d'iceluy il pourra boire de l'oxymel faict cōme s'ēfuit. Vous prendrez la quantité de la meilleur eau que pourrez recouurer: & pour fix ℔ d'eau y met trez quartre onces de miel, & le ferez bouillir, en l'efcumant iufques à la confomption de la troifiemie partie: puis fera coulé, & mis en quelque vaiffeau de verre: puis on adioustera trois ou quatre onces de vinaigre, & fera aromatifé de canelle fine. Pareillemēt pourra vfer de l'hypocras d'eau fait en cefte forte. Prenez vne quarte d'eau de fontaine, fix onces de fuccre, deux drachmes de canelle, & le tout enfemble coulerez par la chauffe d'hypocras, fans aucunement le faire bouillir...*"

A much later French language text *Avis de Precaution* [1721] has among its thirst quenchers a tamarind tisane (like lemons, tamarind must presumably have been an somewhat expensive item to source), but there is also a cheaper rose water for parched tongues and a vinegar gargle: "*Contre la foif de l'eau d'orge, & du jus, ou du firop de limon, de la ptifane aux tamarinds &c. Pour humecter la féchereffe de la langue, de l'eau rofe, & du vinaigre en gargarifme, ceux qui craignent l'odeur de l'eau rofe mettront de l'eau de pourpier, ou d'ofeille.*"

Diet overall has been touched upon elsewhere but here the emphasis is on food used almost as medicinal cuisine, particularly in one of the later French passages that follows. Sometimes food was viewed as a general 'preservative' against plague while in others the diet was adjusted to fit in with the treatment regime prescribed by physicians as they sought to balance the Humours of their patient. When Thayre's *Excellent Treatise* [1625] was published the educated section of the public for whom the book is written were living through a plague outbreak, and he recommended the following: "*And it is now exceeding good with all your meats to vfe fharp fauce made of vinegar, or rofe vinegar, Orenges, Limons, Pomegranates, and a little Cinamon and Maces, but forbeare and refufe all hot fpices, and ftrong wines, Onions, Garlicke, Leekes, Cabage, Radifh, Rocket, and fuch like, the vfe of them is very hurtful and dangerous. But thefe are good and wholfome, Borage, buglosse, sorrell, endiue, cichory, violets, spinage, betony, egrimony* [sic], *they are good both in salades, sauces, and broth, and your dyet ought in this time of infection to be cooling and druing* [sic].

The vse of Orenges, limons, pomgranats, is very good; so is Vinegar, Cloues, maces, saffron, sorrell with your meat, or either of them in a morning with sugar is good. Let all your meates bee drest and saust with Vinegar, Orenges, and Limons, Maces and Saffron, and a little Cinnamon, and auoide all strong wines, and hot spices."

Returning to Royet [1583], about four decades prior to Thayre, the following passage says 'potage herbs are more medicines than food' in the opening sentence which shows how closely food was allied to treatment, at least in French eyes:

"*Les herbes potagieres seruent plus de medicamens que d'aliment. Dauantage le ius, ou decoction d'icelles est plus saine que ne sont les mesmes herbes: aussi n'est besoin d'vser trop souuent de telle viande, si la necessité ne le porte: car elles engendrent des humeurs; creus, & aqueuses, lesquelles se pourrissent aisement. L'on choisira donc les aigres, ou quelque peut ameres: celles aussi qui auront puissance de deseicher mediocrement: toutesfois si le corps est cholerique, & la chaleur grande, l'on pourra vser d'herbes humectantes. Les crues sont lesplus dangereuses: pourautant que le ventricule ne les peut cuire facilement.*

A raison de quoy les salades serort beaucoup meilleures, si l'on fait bouillir la laictue, la cichoree, ou l'endiue, l'oseille, le pourpier, & autres semblables herbes, desquelles on fait ordinairement salades. Si l'on craint la froideur de la laictue, on pourra mesler du basilic parmy, ou de la mente: Et poursaire sallade bonne à toutes gens, faut laisser l'huyle, & tēperer l'aigreue du vinaigre auec raisins de Corynthe, & puis y ietter force sel par dessus, & les manger en esté à midy."

In two other Royet snippets we have the following:

"*Des herbes potagieres.*
L'on peut bien vser des potaiges de borrache, & buglosse en toute saison, & aussi de toutes ces especes de cichoree, d'ēdiue, d'oseille, & de la petite pimpenelle, de laquelle on se sert volontiers de cōtrepoison, en quelque sorts que l'on en veuille vser.

Si le malade est fort debile, on luy donnera de la gelee faite de chapon, & veua, y faisant bouillir eau d'ozeille, de chardon benit, borrache, & vn peu de vinaigre rosat, canelle, succre, & autres choses qu'on verra estre necessaires."

That last paragraph very much follows the 1649 English translation of Paré's work which was published in 1575, and one wonders if Royet sourced from Paré or whether medicinal cuisine was quite commonplace in France. Paré comments: "*If the patient bee somwhat weak, hee must bee fed with Gellis made of the flesh of a Capon, and Veal sodden together in the water of Sorrel,* Carduus Benedictus, *with a little quan-* [sic] *of Rose-vineger, Cinnamon, Sugar, and other such like, as the present necessitie shall seem to require.*"

Another piece from Paré recommended: *"Eggs potched and eaten with the juice of Sorrel, are verie good. Likewiſe Barlie-water ſeaſoned with the grains of a tart Pomgranat, and if the Fever bee vehement, with the ſeeds of white Poppie."*

A more in-depth view of French medicinal cuisine is to be seen in this later 18th century text, *Avis de Precaution* [1721]: *"Les Peſtiferez avec fiévre, ſoit celle qui accompagne la ſortie des bubons, laquelle ne doit durer qu'un jours, ſoit fiévre ētenduë ou compliqueé avec les accidens pourriture, ceux-là ſe contenteront de boüillons chargez de quelques plantes, comme oſeille, ſcabieuſe, pimpinelle, bourache, jus de citron, &c. leur boiſſon ſera de l'eau de poulet, ou de l'eau panée, ou de la ptiſane, avec une once de racines de ſcorſonére ou d'oſeille ou d'agrimoine, deux pincées d'orge entier, autant de raſins ſecs, demi citron coupé par tranches, avec un peu de ſucre pour une pinte: On pourra y ajoûter quelquefois au lieu de citrons des tamarins, d'autres fois une dragme de nitre épurée par criſtalliſation, & on obligera le malades de boire largement."* Note at the end there how sour-tasting tamarinds are seen as a substitute for citrus.

In *La Pratique de Medecine* [1682] Rivere suggests that an alternative to juleps is a regime of medicinal cuisine; the ingredient list including some of the familiar plants: *"En place des juleps pour les plus delicats, l'on peut quelquefois ſubſtituer, des boüillons medicinaux, alterez des feüilles de bourrache, d'ozeille, de pimpinelle, & des autres herbes plus agreables a goût, des écorces de pommes de bonne odeur, de la poulpe, & aigre du citron, aven un poulet, & l'on y peut quelquefois ajoûter le ſel de prunelle pour rafraichir davantage."*

While physicians may have concocted potions for customers lucky enough to afford the steep fees there were some self-help and home remedy treatises available. These were ideal for households unable to afford medicines, the services of a local physician, or simply lived too far away from medical help and depended upon personal self-reliance in desperate times. These self-help books, however, rather assumed you possessed the ability to read, something the very poorest in society would not have able to do.

One of the most widely published and read domestic self-reliance handbooks during the 17th century was that written by Gervase Markham, who wasn't past bundling his own work with reprints of other writer's work. Generally sold as *The English Housewife*, or sometimes combined in *A Way to Get Wealth*, the books were reprinted numerous times and were a sort of 17th century Mrs. Beaton. The following passage is from a 1668 copy and represents the kind of household knowledge that would probably have been widely followed in daily life:

"For the Peſtilent Feaver which is a continuall ſickneſs full of infection and mortality, you ſhall cauſe the party firſt to be let blood, if his ſtrength will bear it: then you ſhall give him cool Julips made of Endive or Succory water: or the ſyrup of Violets, conſerve of Barberies, and the juyce of Lemons well mixed and ſymbolized together.

Alſo you ſhall give him to drink Almond-milk, made with the decoction of cool herbs, as Violet leaves, ſtrawberry leaves, french mallowes, purſlane, and ſuch like; and if the parties mouth ſhall through the heat of his ſtomack, or liver, inflame, or grow ſore, you ſhall waſh it with the ſyrup of Mulberies, and that will not only heal it, but alſo ſtrengthen his ſtomack. If (as it is moſt common in this ſickneſs) the party ſhall grow coſtive, you ſhall give him a ſuppoſitory made of honey, boyld to the height of hardneſs, which you ſhall know by cooling a drop thereof, and ſo if you find it hard, you ſhall then know that the honey is boyl'd ſufficiently; then put ſalt to it, and ſo put it in water and work it into a roul in manner of ſuppoſitory, and adminiſter it, and it moſt aſſuredly bringeth no hurt, but eaſe to the party, of what age or ſtrength ſoever he be: during his ſickneſs you ſhall keep him from all manner of ſtrong drinks, or hot ſpices, and then there is no doubt of his recovery."

Another section of Markham's work provides readers with instructions for self-preservation from the plague, and also what to do if struck with the disease:

"To preſerve your body from the infection of the plague, you ſhall take a quart of old ale, and after it hath riſen upon the fire, and hath been ſcummed, you ſhall put thereinto of Ariſtolochia longa, *of* Angelica, *and of* Celandine, *of each half a handful, and boyl them well therein; then ſtrain the drink through a clean cloth, and diſſolve therein a dram of beſt* Mithridate, *as much* Ivory *finely powdered and ſearſt, and ſix ſpoonfuls of* Dragon *water, then put it up in a cloſe glaſs; and every morning faſting, take five ſpoonfuls thereof, and after bite and chaw in your mouth, the dried root of* Angelica, *or ſmell on a noſegay, made of the taſſeld end of a ſhip rope, and they will ſurely preſerve you from infection.*

But if you be infected with the plague, and feel the aſſured ſigns thereof, as pain in the head, drought, burning, weakneſſe of ſtomack, and ſuch like: Then you ſhall take a dram of beſt Mithridate, *and diſſolve it in three or four ſpoonfulls of* Dragon *water, and immediately drink it off, and then with hot cloaths or bricks made extream hot, and laid to the ſoles of your feet (after you have been wrapt in woollen cloths,) compel your ſelf to ſweat, which if you do, keep your ſelf moderately therein, till the ſore begin to riſe; then to the ſame apply a live pidgeon cut in two parts, or elſe a plaiſter made of the yolk of an egg, hony [sic], herb of grace chopt exceedingly ſmall, and wheat flower [sic], which in very ſhort ſpace will not only ripen, but alſo break the ſame without any other inciſion; then after it hath run a day or two, you ſhall apply a plaiſter of* Melilot *unto it, untill [sic] it be whole."*
That mention in the first paragraph above, of smelling on the tassled end of a ship's rope, is presumably recommended because of the aromatic tars that would possibly have been used to preserve the ropes. Herb of grace in paragraph two is the plant rue, widely used in many anti-pestilential remedies.

Medical professionals had their own means of self-preservation as they went about the daily business of tending to the plague-sick. For example, in *Regimen Contra Pestilentiam* [1483] there is the following 'tip': *"... therfore brede or a ſponge ſopped in vynegre I take with me holdyng it to my mouth and noſe be cauſe alle egre thynges ftoppen the wayes of humours / & fufficiently no venemous thyng to enter in to a mannys body and fo I efcaped the Peftilence my fellowes fuppofing that I fholde not lyve.*"

Another contemporary work [1485] advises that: "... *Wyfe Phyficyens in vifityng feke folke ftande ferre from the pacient holdyng their face toward the doore or wyndowe and fo foolde* [sic] *the feruants of feke folke ftande.*" The word 'foolde' there is *should*.

About three hundred years later De Mertens in his account of the 1771 Moscow plague says this about his precautions on household visits: "*When I looked at a patient's tongue, I ufed to hold before my mouth and nofe a pocket-handkerchief moiftened with vinegar.*"

* *Although the atmofphere may not be capable of communicating the peftilential contagion beyond a very limited diftance from its fource, yet to approach fo near as within a foot of the infected, appears to us to be a practice not generally fafe.*"

In time of pestilence households might make up their own preventative potion or medicine and take this daily in the hope of fending off a plague attack. Some of these concoctions had a very complex set of ingredients which would need to be sourced from an apothecary while others were far more rudimentary, often using ingredients that might be found in fields or hedgerows. In his *Dialogue* (1578), which was written for the lay-person, Bullein gives the following recipe:

"*A good powder:*
One better, I assure you, then a kinges raunsome, and thus it must be made: take the leaues of Dictamnus, *and the rootes of Tormentill, of Pimpernell, of Seduall, of Gentian, of Betonie, of eche halfe an unce; bole Armoniacke, prepared, an vnce; Terra Sigellata, iij dragmes; fine Aloes & Myrrhe, of eche half an vnce; Safron, a dragme; Masticke, ij dragmes: beate them together finely and searsed. This is the pouder: of this must a dragme be dronke in iiij or vi sponfull of Rose or Sorell water, when danger approacheth, or in the tyme of danger.*"

The remedy above was regarded as a strong one so for those with weaker stomachs Bullein suggested: "*The sirrupes of Violettes, of Sorell, of Endiue, of sower Limondes, of eche like, mingled with Burrage water, and a Ptisane made of Barlie mingled together, is verie holsome to drinke: put in the pouder of bole Armoniack, whiche is of a singular vertue to coole; for Galen did help thousandes at Rome with the same Bole and the* Theriaca *mingled together, in a greate pestilence.*"

Pomanders became an accessory, not so much for fashion and decoration but for their perceived ability to keep evil infections away. Thayre [1625] has this:

"*Another good Pomander, thought not all thing fo coftly to bee worne againft the infection of the ayre.*

R. Of the rinds of Citrons one dramme; Storax, Calamint, two drammes, Labdanum one dramme, of all three kinds of Sanders, ana two scruples; flowers of Rofes, Violets, and Nenuphar ana halfe a dramme; liquid Storax, Beniamin, ana one dramme; Camphor one fcruple, Muske and Ambergreece ana three graines, with Rofe-water, & gumme Dragagant a little quantitie make your Pomander."

While orange or citrus peel found its way into pomanders it was also one of the recommended substances to be chewed as a prophylactic, Bradwell [1636] suggesting: *"Carrie in your mouth a peece of* Citron-pill, *or for want of that, of* Lemon-pill; Clove, *or* a peece of Tormentill Root."

During the Great Plague of London the Royal College of Physicians [1665] suggested further herbal items to carry with you when going out and about, or to be transformed into a pomander as follows:

"Such as are to go abroad, fhall do well to carry Rue, Angelica, Mafterwort, Myrrhe, Scordium, or Water-germander, Wormwood, Valerian, or Setwall-root, Virginian-fnake-root, or Zedoarie in their hands to fmell to...

Or they may ufe this Pomander.
Take Angelica, Rue, Zedoarie, of each half a dram, Myrrhe two drams, Camphire fix grains, Wax and Labdanum *of each two drams, more or lefs, as fhall be thought fit to mix with the other things; make hereof a ball to carry about you; you may eafily make a hole in it, and fo wear it about your neck with a ftring."*

The pomander recipe above had appeared in a previous 1636 College booklet on the plague. Picking up again on Thayre [1625], his book provided his readers with a wealth of preservative alternatives to pomanders:

"Another preservative:
Three or foure graines of Bezoar ftone taken in the morning in a fpoonefull of fcabious, or Sorrell water, is a good preferuatiue....
Alfo to vfe the roote of Angelica, fteeped in Vinegar to chew in your mouth as you go in the ftreete is good, and to eate a little thereof.
Gentian, Zedoary, Turmentill, chewed and kept in your mouth are good.
Sorrell eaten in the morning with a little good Vinegar like a fallet, is very good: the vfe of Orenges and Lemons is very good, Pomegranates and Vinegar.
It is good euery morning betime, to take fome good preferuatiue, and before you go abroad, it fhall not bee amiffe to eate fomething to your breake-faft that is wholefome, as bread and fweete butter, a potcht egge with Vinegar, or fome other thing as you are prouided, and vie alway in going into any infected place a roote of Angelica to chew vppon in your mouth, a little fpunge dipt in Rofe Vinegar to fmell vnto oftentimes is good, put into a Pomander box of luory [sic].

Preferuatiues for the Commons and Country-men, who haue not an Apothecary at hand.

Take of Rue or Hearbe-grace two ounces, of the young buds of Angelica two ounces; or for want thereof, of the root or feed one ounce, Bolearmoniacke prepared one ounce, of Iuniper berries one ounce, of Walnuts cleane picked from their skins, two ounces, good figs in number fix or feauen, of Saffron fix pennyworth, of good wine vinegar that is fharpe foure ounces. Let thefe be well beaten together in a mortar, the fpace of one houre, and then put in your Vinegar, and incorporate them together. Which being done, put it into fome fweet gally pot or glaffe, and couer it clofe; and take thereof daily in the morning the quantity of a Nutmeg. Or you may eate thereof at any time going neere, or in any infectious place.

Another good preferuatiue of no leffe vertue in refifting all infection.
R. Of Holy Thiftle, or for want thereof, our Ladies Thiftle fo called, Beroni, Angelica, Scabious, Sorrell, Pimpernell, Turmentill, of either of thefe a handfull, Gentian rootes alfo, if they may be had.

Bruife all thefe in a ftone Mortar a little, and put thereto a pinte of good Vinegar, and halfe a pinte of white Wine, and put them into a Still, and draw forth the water, and take two or three fpoonfuls thereof euery morning fafting, and be free from all infection.

The roote of Angelica *layd or fteeped in good Vinegar all night, and a little thereof taken in the morning is a good preferuatiue. The feeds are of the like vertue."*

While pomanders might provide the wearer with a fragrant scent rather than any particular prophylactic advantage there were also physicians who advocated wearing medicinal amulets. In *London Practice of Physic* [1685] Willis says amulets were to be worn round the neck or wrists and '... *are thought to have a wonderful Force againft the Peftilence*' with the ingredient list sometimes consisting of arsenic and mercury, but also '... *the Powder of Toads, and other venemous Things.*' All of which rings bells in the department of quackery. That said, the English physician George Thomson published a book in 1666 [*Loimotomia*] in which he describes his dissection of a plague corpse the previous year, and protecting himself from this hazardous enterprise (something akin to undertaking the autopsy of an Ebola victim without any protections) with a dried toad hung round his neck!

However, in the 'preservative' battle against the plague, potions and fluid medicinal liquids to fortify the body and Spirits were common, often taking the form of 'waters' and vinegars. In the case of the following recipe from Willis [1691] the pestilential vinegar was taken on its own or mixed with posset:

"For fome that are of a hotter Conftitution, and a high Sanguine Temper, it may be proper to take every Morning a fpoonful of Peftilential Vinegar *in a little* Carduus water, *or plain* Wallnut *water; or elfe drink a draught of* Poffet-Drink, *made with a Spoonful of that Vinegar.*

Let the Peſtilential Vinegar *be made thus.*

Take of the Roots of Angelica, Butter-Burr, Tormentil, Elecampane, *of each half an Ounce,* Virginian - Snake-*Weed, choice* Zedoary, Contrayerva, *of each three Drams;* Leaves of Scordium, Rue, Goats-Rue, *of each one handful;* Marigold Flowers, Clove-gilloflowers, *of each half a handful; Seeds of* Citron and Carduus, *of each two Drams; Cut and bruiſe theſe, and put them in a Glaſs-Bottle, with three Pints of the beſt* Vinegar, *to digeſt for ten days."*

In addition to anti-pestilential vinegars there were 'waters' of all sorts, the one below from a 16th century French work, *Tresor des Remedes Secretz* [1557]. The recipe has an enormous number of ingredients, some which would have been expensive in their own right even before they had been concocted into a potion. The work from which the recipe is taken is essentially a book on alchemy, chemistry and distillation somewhat reminiscent of Hieronymus Brunschweig's *Art of Distillation* [1500], and draws together remedies from other attributed medical sources, although the one below is unattributed:

"Eau de vie contre la peste
PRENEZ, Rue, Sauge, fleurs de Lauēde, Mariolaine Abſynthe, Romarin, Roſes rouges, Chardon benedict, Pimpenelle, Tormentille, Valeriane, grains de Geneure, Bayes de Laurier, Terre ſigillee, Bole Armenic preparé, ań. deux dragmes. Dictamne, Angelique, Biſtorte, eſcorce de Citron, Meliſſe, Gedoar [sic], Campane, Gentiane, Rhapontic. ań. trois dragmes. Coriandre preparee, fleur de Borrache, Bugloſe. Sandal blanc & rouge, ſemence de Vinette, Baſilic, Rheubarbe, Ben blanc & rouge, Graine de Paradis, Poiure. ań. dragme & demie. Gingembre deux dragmes, Cinnamomme, Safran, eſpeces de confection contre peſte, Electuaire deliurant, Electuaire de Gemmes, Diamoſchi doux, Diacameron, Diambre, diarhodon Abbatis, & du laetifiant d'Almanſor. ań. vne dragme. Calame Aromatic, Girofle, Macis, Noix muſcade, Cubebes, Cardamom, Galange, Agalloch, ou bois de Aloës ań. deux ſcrupules. Os de cœur de Cerf, Spic de nard, Camphre, ań. demie dragme, huyct fueilles d'Or, demy ſcrupule de Muſch, De bon Theriacle eleu quatre onces, de bon Methridat deux Onces, vin ſublime, & rectifié, deux meſures. Tout cela ſoit deſtillé en Alembic."

Angelica makes a frequent appearance across both the prophylactics and remedies used in time of plague, and it was commonly used as a personal street-level preservative, being chewed in the mouth as physician Thomas Coghan tells us in *Haven of Health* [1636]: *"Angelica is hot and dry almoſt in the third degree. It is a rare herbe, and of ſingular vertue, but chiefly commended againſt the Peſtilence, as well to preſerve a man from it, as to helpe him when he is infected... And ſo was I wont to uſe it at Oxford in time of Plague, to grate of the dry root into drinke, and to carry a little piece of the root in my mouth when I went abroad."*

Elsewhere Coghan elaborates with further advice on measures to be taken before stepping out of the house: *"But now to arme the heart againſt this infection, when you have occafion to goe forth of the houfe, having firft eaten or drunken fome what, for it is not good to goe forth with empty veines or elfe having received a fume... you fhall put in your mouth a Clove or two, or a little Cinamome, or a peece of Setwall, or of an Orenge pill, or beft of all, a peece of the roote of* Angelica, *or* Elecampane, *and take in your hand an Orenge, or a pofie of Rew, or Mynt; or Balme: Or elfe carry with you a handkerchiefe or fpunge* [sic] *drenched with white Vineger* [sic] *of Rofes, if you can get it, if not common Vineger* [sic]*, efpecially white."*

Not all medics appear to have believed chewing on angelica and other items was a wise course of action for those working in the medical profession. In the 18th century Heister [1750] warns his fellow practitioners: *"Being come to the Patient's Apartment, great Care muft be always taken that we* neither eat nor drink there, nor even fwallow our Spittle. *For there is no fmall Danger in that Cafe of fwallowing the volatile peftilential Exhalations or Effluvia, by which means our internal* Vifcera *and Blood would be infected: For which Reafon we cannot approve of the Cuftom of fome who are continually chewing and fwallowing Myrrh, Cinnamon, Angelica, Zedoary, or the like, all the time they are in an infected Place. For as fuch Things excite a plentiful difcharge of* Saliva *into the Mouth, it is hardly poffible but fome of the infectious Effluvia will be intangled therein, and fo go down into the Stomach, and get into the Blood. But the chewing of fuch Aromatics may be very proper at home, as they are in their own nature wholefome; the ufe of them in the former Cafe being improper only as to Time and Place."*

While Coghan grated angelica root into preservative drinks the Royal College of Physicians of London [1665] suggested a further way of using angelica root in grated form: *"Take the root of Angelica beaten grofly, the weight of fix pence, of Rue, and Wormwood, of each the weight of four pence, Setwall the weight of three pence; bruife thefe, then fteep them in a little Wine-Vineger, tye them in a linnen cloth, which they may carry in their hands, or put it into a Juniper box full of holes to fmell to."*

Willis [1691] also has a preparative variation on angelica root, re-drying it after macerating the root in vinegar: *"For Medicines to hold in the Mouth, and chew on, Roots of* Zedoary, Contrayerva, *and* Snakeweed *are very good; alfo Roots of* Enula-campane, Angelica, *and* Mafterwort; *thefe either alone, or macerated in* Vinegar, *and dried again.* Myrrh *is very excellent: Some commend* Tobacco, *and chew it almoft continually."*

From the following century is the next recipe from *Avis de Precaution* [1721] for prophylactic angelica pills to be carried in a box and chewed as needed:

"Paftilles béfoardiques à mâcher.
Prenez de la racine d'angèlique, & de zedoaire, de l'ècorce de citron fèche de chacune deux onces, du maftic une once, du canfre une dragme, mettez en poudre, & incorporez avec du mucilage de gomme adragant fonduë au vinaigre fimple, ou diftillée, pour

former de petites paftilles, à porter fur foy dans une boëtte, pour en tenir toûjours à la bouche." Note, however, in the recipe above there is no actual bezoardic stone ingredient despite the title header.

From precisely the same period Joseph Browne [1720] refers to a recipe from a French painter, a M. Mahew, who had survived four plague outbreaks using the following means of self-preservation: *"Take Galingal fliced 3ij. infufe this in a Quart of White-Wine Vinegar for three Days; of this take four Ounces every Morning, in which fteep a Toaft of Bread, and eat it; fnuff fome of the Vinegar up your Noftrils, and wafh your Hands and Temples with the reft. About his Neck and Arms, he wore Amulets of Amber, Camphore and Euphorbium fewed in Bags, and bound about the Parts."*

In a further twist on toasted bread Browne also includes this recommendation for a preventative: *"Or, elfe take Rue one handful, ftamp it in a Mortar; put thereto Wine Vinegar enough to moiften it, mix them well, then ftrain out the Juice; wet either a piece of Spunge* [sic] *or a Toaft of brown Bread therewith, to carry about with you."*

Browne also presents some recipes for amulets and pomanders, among them a pomander for rich clients: *"Take Citron Peel, Angelica Seeds, Zedoary, red Rofe leaves, of each 3fs. Yellow Sanders, Aloes Wood an. Ɔi. Nutmegs Ɔiv. Storax, Benjamin, of each 3i. Camphore gr. vi. Labdanum 3iij. Gum Tragacanth diffolved in Rofe-water enough to form it. To this add of Spirit of Rofes fix Drops, (or the fame Quantity of Bergamot) put this in a Box, or wear it about the Neck."*

Depending on whether you were rich or poor your physician would adjust or change the remedy to suit your wallet, although some physicians would treat the poor for nothing, subsidizing the cheaper medicines through fees paid by their rich clients. In some cases wealthy clients would have powdered gemstones and semi-precious stones added to their potions. Quite what medicinal value these additional items could add to a medicinal brew has to be somewhat doubtful, though one supposes that by spending enormous amounts of money the recipient of the potion would feel as if at least they were paying for health and cure.

Differentiation between remedies and treatment for poor or wealthy are typically exemplified in Textor's *De la maniere de Preserve de la Pestilence* [1551] where the ingredient list for a *'Poudre cordiale contre le uenin de la pefte, pour les riches & puiffans, & pour les princes'* contains sapphires, *'Scobis eboris'* (ivory dust), and *'Cornu monocerotis vulgò vnicornu'*, or the horn of the knarwal. Quite how precious that ingredient was is perhaps anyone's guess. Meanwhile the following recipes are then graded according to the patient's pocket: *'Pour ceux qui font de fortune mediocre autre poudre excellente'* and *'Pour le poures poudre qui neft pas de moindre efficace que la seconde'*.

Let us turn to one of the best known English herbal sources for an example of a plague remedy that cost heaps of money to buy, Nicholas Culpeper. The two examples below are really high-priced medicines; not only containing individually expensive ingredients but also requiring a huge investment of time in the making of the potions. The text extracts come from a 1720 edition of Culpeper's *Pharmacopoeia Londinensis, or the London Dispensatory* originally published in 1653. That particular work has a highly significant place in British medico-herbal history as it was the first time the Latin dispensatory of the London College of Physicians had been translated into English. Prior to that Physicians could hide their remedies behind Latin phrases and words that were a mystery to a large portion of the population, thereby keeping the public in ignorance while physicians could profit from their closed-shop knowledge. Culpeper's translation essentially broke the monopoly of physicians on providing medicines, allowing the apothecaries' trade to intervene in the remedy supply chain.

You will find the text extracts below sectioned into 'College' and 'Culpeper' parts; the College section shows recipes from the College Dispensatory while the Culpeper part represents his personal comments on the remedy. Sometimes his assessments were negative, sometimes positive. The first one is an elaborate alcoholic cordial 'water', *Aqua Celeſtis*, attributed to Mathiolus:

"*College. Take of Cinnamon an ounce, Ginger half an ounce, white, red and yellow Sanders, of each ſix drams; Cloves, Galanga, Nutmegs, of each two drams and an half; Mace, Cubebs, of each one dram; both ſorts of Cardamoms, Nigella ſeeds, of each three drams; Zedoary half an ounce, ſeeds of Aniſe, ſweet Fennel, wild Parſnips, Bazil [sic], of each a dram and an half; Roots of Angelica, Avens, Calamus Aromaticus, Liquorice, Valerian the leſs, the Leaves of Clary, Time, Marjoram, of each two drams; the Flowers of red Roſes, Sage and Roſemary, Betony, Staechas, Bugloſs, Borrage, of each one dram and an half: Citron peels three drams: Let the things be bruiſed that are to be bruiſed, and infuſed fifteen days in 12 pints of the beſt ſpirit of wine, in a glaſs bottle well ſtopped, and then let it be diſtilled in Balneo Maria according to Art. Adding to the diſtilled water, Powders of Diambra[1], Diamoſchu dulce[2], Aromaticum, Roſatum, Diamargaritum frigidum[3], Diarrhodon Abbatis[4], Powder Electuarti De Gemmis, of each 3 drams: Yellow Sanders bruiſed 2 drams: Musk, Ambergreeſe, of each a Scruple tied up in a fine rag, clear Julep of Roſes a pound, ſhake them well together, ſtopping the glaſs cloſe with wax and parchment, till it grow clear, to be kept for your uſe.*

Culpeper. It comforteth and cheriſheth the heart, reviveth drooping ſpirits, prevaileth againſt the Plague and all malignant Fevers, preſerveth the ſenſes, and reſtoreth ſuch as are in conſumptions. It is of hot nature. Let not the quantity taken at a time exceed half a dram.

Only take this caution, both concerning this, and all other ſtrong waters. They are not ſafely given by themſelves in Fevers, (becauſe by their hot quality they inflame the blood, and add fuel to the fire) but mixed with other convenient Cordials, and confideration had to the ftrength, complexion, habit, age, and fex of the Patient."

[1] *Diambra* is a powder mixture that includes cinnamon, angelica root, cloves, mace, nutmegs, galangal, Indian spikenard, ambergrease, and many other spice ingredients. Culpeper in his notes on *Diambra* says that Mesue used this for the Head where the mixture heated and strengthened the brain, caused 'mirth', and strengthened the heart. The powder was sometimes mixed with honey to make an electuary.

[2] *Diamoschu dulce* was another herbal 'powder' and contained saffron and other spices, wood of aloes, pearls, toasted raw silk, coral and amber. Culpeper says it '... *wonderfully helps cold afflictions of the Brain that comes without a Fever.*' Again, this could be made into an electuary.

[3] *Diamargaritum frigidum* is a convoluted powder mixture that contained the seeds of purslain, white poppies, endive, sorrel and citron, plus various spices, aloe wood, various flowers, ivory, pearls and camphor. Culpeper is not enthusiastic about this remedy, regarding it as costly.

[4] *Diarrhodon Abbatis* was a 'powder' with the ingredient list including sanders, gum tragacanth and arabic, ivory, asarabacca roots, mastic, Indian spikenard, cardamons, liquroice, saffron, aloe wood, cloves, anise and fennel seeds, pearls and camphor among many others.

In this second recipe, for *Species Confectionis Liberantis*, we find that while one or two simple vegetable ingredients are present so also are emeralds, pearls, amber, ivory and silk, with Culpeper commenting that musk and ambergrease have been left out because of cost:

"*College. Take of Tormentil roots, the feeds of Sorrel, Endive, Coriander prepared, Citron, or each one dram & half; all the Sanders, white Dittany, of each a dram; Bole armenick, Earth of Lemnos, of each 3 drams, Pearls, both forts of Coral, white Amber, Ivory, Spodium, bone of a Stags heart, the roots of Serpentary, Avens, Angelica, Cardamoms, Cinnamon, Mace Wood of Aloes, Caffia Lignea, Saffron, Zedoary, of each half a dram; Penids, raw Silk roafted, Emeralds, Jacinth, Granate, the flowers of water Lillies, Buglofs and Red Rofes, of each one fcruple; Camphire feven grains; make them into powder according to art.*

Culpeper. The Serpentary roots are added, and Musk and Ambergreefe, of each 3 grains left out; becaufe deftructive to the Common wealth. It is exceeding good in peftilential Fevers, and preferves from ill airs, and keepeth the humours in the body from corruption, it cools the heart and blood, ftrengthens fuch as are oppreffed by heat. To conclude, it is a gallant cool Cordial, though coftly. It being out of the reach of a vulgae [sic] *mans purfe.*"

Quite why medics thought that precious stones would improve the potency of a remedy one wonders, although Bullein in his *Bulwarke* [1579] puts the use of sapphires down to *Dioscorides*, a Greek physician working in the first century AD: *"Dioscorides affirmeth, this heauenly coloured precious ſtone, is wholſome to be drunke againſt the ſtinging of Serpents, to reſyſt the poyſon: and being drunke it helpeth all exulcerations of the Guttes..."*

Treatment regimes for poor folk were pretty rudimentary and were composed of very simple items such as wine, vinegar and garden herbs. From Textor [1551] is the following list of fifteen cheap plague remedies for the poor:

"Antidots ou medicaments preſeruatifs & confortatifs de uile prys, faciles à trouuer & à preparer pour les poures.
1 Prenez dun ail & buuez vu peu de vin pur apres.
2 Ou dune figue auec vne noix & de rue & vn peu de ſel meſmes en hyuer.
3 Ou vingt fueilles de rue auec deux noix, autant de figues & vn grain de ſel, le tout meſlé enſemble pour prendre incontinent au matin.
4 Ou ſix fueilles de rue auec vinaigre.
5 Ou la racine de lherbe appellee vulgairement auſtruche en latin imperatoria, daucūs laſerpitiū Gallicū.
6 Ou la racine dangelica.
7 Ou de gentiana.
8 Ou de Zedoar.
9 Ou de chardon benit.
10 Ou de carline appelle daucuns le bon chardon.
11 Ou de lherbe nommee ſcordium de lune, de deux, ou de pluſieurs en pouldre bien menue à la quantité dune drachme, ou en maſſe bien molle faite de miel cuit & de vinaigre, ou de quelque ſyrop propre, comme de limons, ou en opiate le gros dune chaſtagne, ou dun pois cice auec du vin en hyuer, en eſté auec eaue roſe ou auec ius doxeille.
12 Oxeille ſeule ou auec pimpinelle trempee en vinaigre pour prendre au matin.
13 Ou le ius dicelles ou de pourchaille auec vn peu de vinaigre: de quoy on pourra faire vne toſtee en eſté.
14 Ou graine de geneurier, fueilles verdes de pimpinelle, de betoine, de puliot, doxeille autant dune que dautre broyees enſemble cuites auec miel cuit & vn peu de vinaigre en façon de conſerue
15 Ou le ius dicelles herbes cuit auec miel ou auec ſucre & vn peu de vinaigre en ſyrop ou en forme plus eſpoiſſe pour en prendre deux culliers ou trois le matin."

Typical among Textor's remedies for the poor is the following, containing a drastically cut-down ingredient list and containing sorrel and vinegar:

"Potion pour le poures.
Suc doxeille bien cler, trois onces.
Doranges aigres
Bon vinaigre blanc & cler

Eaue rofe de chacune chofe vne once. Faites bruuage, en adiouftant vn peu de fuccre qui voudra."

Meanwhile *princes*, the *puiffans* and *mediocres* would get an ingredient uplift to suit their pockets, and to include *succi limonis, vini mali punici*, white wine and rose water, and in another potion *syrupi granatorum* (pomegranate), mixed in with lemon, white wine, rose water, sorrel juice, sharp oranges, and vinegar.

In a section on *Des remedes prouocatifs de la fueur: & des uomitoires* Textor has a potion for '*emouuoir la fueur*' best suited to those with deeper pockets, which contained theriaca or mithridat bonae, bol armeniac clay, and scabious water. This is followed by a simple potion for the poor that included those very simple garden herbs parsley and anise, but also red poppy flowers:

"Autre potion facile comme pour les poures.
Decoction de fcabieufe & de fleurs de pauot rouge, vn voirre auec vn peu de fuccre.
On ptifaine faite dorge & de femence danis & de racine de perfil. Ou decoction de poix cices, de racine de perfil, de racine de cicorre: lune ou lautre decoction auec fyrop aceteux deux onces, ou auec fuccre & vn peu de vin aigre blanc. Ces bruuages fe douuent bailler chauds au malade eftant bien couuert au lict."

From a similar time frame Royet [1583] outlines a number of 'opiate' remedies. However these have nothing to do with opium and it appears that perhaps the use of the word opiate is down to a more general meaning of sedation, dulling, or deadening. Many of the Royet recipes use expensive spices but there is this one for the poor and needy incorporating dried nuts:

"Vne autre pour les poures
Rx. Conferuae rofarum, Enulae camp. an. ℥j.
Rd.ireos, ℥. fs. Nucum aridarum non rancidarum, Foliarum ruthae, an.3. ij. Sem.citri, vel arantij, Hippericonis, Baccharũ iuniperi, an.3. i. Succi oxalidis, & Bugloffæ, an.q. fufficit. Cum melle rofato collato, fiat Oppiata."

The simplicity with which cheap remedies could be concocted can be seen in Thomas Sherwood's *Charitable Pestmaster* [1641] in which he comments about the expense of a certain 'vomit' that he suggests, then goes on to say: *"But the poorer fort that cannot go to this charge, may take inftead thereof Aloes one dram in the pap of an Apple, ftewed Prunes, or elfe in a little Ale or Beere... After the bodie is purged, it fhall be neceffary to draw fix or eight ounce of bloud from the liver or middle vein of the arme, if the partie be able to fuftain the loffe of it."* Blood-letting, it seems, was never very far away from any treatment regime.

The French, with their long tradition of wine making, also put that particular product to good use in the treatment of poor people during plague outbreaks. Boghurst [1666] too, was a cheerleader for wine as a cordial in plague times, but in the French text, *Medicine et Chirurgien Des Pauvres* [1674], the anonymous author says: *"Le vin eſt un merveilleux preſervatif pour les pauvres, que vous rendrez ſpecifique, ſi vous prenez racines d'Angelique & Scorſonere, avec l'écorce de Citron en poudre, que paſſerez avec de bon vin pour en uſer chaque jour un demy verre à jeun."*

From the late 18th century onwards one has the feeling that chemical pharmacy begins to outweigh the strictly herbal approach that had been taken to treat plague down the centuries. However, in the De Mertens [1799] account of the 1771 Moscow plague outbreak it is evident that physicians and chemists were still struggling to find a cure for plague despite some scientific advances: *"But the cure of the plague by the mineral acids and Peruvian bark, is only to be expected when the diſeaſe appears under its leſs violent forms. In a great number of inſtances (where the diſeaſe has been more violent) theſe remedies have been preſcribed, not only without effecting a cure, but even without retarding death for a moment. Various other medicines, ſuch as theriaca (which has been ſo improperly cried up in the plague) camphor, dulcified ſpirit of nitre, &c. have in like manner failed; ſo that we are compelled to acknowledge, that the plague (under its more violent forms) is of ſuch a malignant nature as not to yield to any medicines with which we are yet acquainted, howſoever well adapted they may, a priori, ſeem to be for getting the better of this diſorder. From analogy and the preceding facts, I am inclined to place more reliance upon the Peruvian bark and acids, given in large doſes, than upon any other remedy; joining with them, to obviate debility, camphor, elixir of vitriol, wine, and bliſters."*

Even in the first half of the 19th century physicians and pharmacologists struggled to find effective new plague remedies and, remember, it would be 1894 before the *Yersinia* bacterium would be identified. Whitney, in *Family Physician* [1835] describes the prospect of inhaling mercury sulphide vapour as part of a post-fever treatment regime, and must have been a far from healthy activity: *"It is stated that sudorifics and mercury always do good; that when the disease is inflammatory and attacks the brain, which is known by the incessant headache and furious delirium from the onset, blood should be taken in a standing posture until fainting is completely produced; and when the perspiration which follows it takes place, it should be encouraged by warm drinks. The patient is then to be immediately salivated in the shortest time possible, and for this purpose the Hindoo method of doing it by cinnabar fumigations should be adopted, as follows: bees-wax is melted and spread over strips of cotton cloth; an equal quantity of cinnabar in powder is spread over the waxed strips, which are then rolled up in the shape of candles; the person to be salivated being seated, a blanket is thrown over him, and the lighted cinnabar candle is placed under the blanket so that he inhales the vapor."*

In reading through many of the remedies and treatments for plague in the previous sections the reader may be left with an overwhelming impression that much of it was based on mumbo-jumbo, and sometimes implausible and far-fetched notions of the potency of certain remedy ingredients. It is also important, however, to remember the knowledge vacuum that had existed for countless centuries and the virtually impregnable socio-religio-political framework that so many past generations lived under.

The fact that there were no real medical 'standards', and that knowledge of *Yersinia* was non-existent, presented opportunities for medical charlatans, quacks and the unscrupulous to peddle false hope and useless remedies to the unwitting. Indeed, there is an undercurrent remonstrating against quackery that surfaces in many plague accounts and medical books over the centuries. In his *Bulwarke* [1579] William Bullein writes some verse regarding the skills of trained physicians while railing against the peddlers of medical mischief. The short extract below gives a flavour of the verse:

> "... *Wyth flattering wordes, and trim tales, glofinges they can tell,*
> *As though in naturall Philofophy, they were feene ful well.*
> *With retrogradation, and Lord of the afcendent,*
> *Plasters, Oyles, Pouders, Salues, and matter defendant.*
> *In feeming to be fkilful, in euery euil malady,*
> *Whether it be moyft, colde, burning, hoat and dry.*
> *Yet neyther read Tagaltius, Marianus, Guido, nor Galen,*
> *Olde Hyppocrates, Diofcorides, Rafis, nor Auicen.*
> *Latine nor Englifhe, little or none, do they reede,*
> *Small is their knowledge, mutch leffe is their fpeede.*
> *Yet lacke they no Brimftone, Quickfiluer, or Litarge,*
> *Oyles groffe and lothfome, to beare out the charge.*
> *They haue Palmeftry, and Charmes, at eche wyghtes defire,*
> *Good ftore of bleßinges, for tothache, and faynct Antonies fire.*
> *If yong Babes through Feuers, wyth cold be fhaken,*
> *Then they fay an euill fpirite, the childe hath taken.*
> *A bad Angell of the ayre, an Elfe, or a witch,*
> *When in deede, deere frende, therebe few fitch* [sic]..."

In the 17th century Nathaniel Hodges [1720] had felt obliged to utter a few words on quacks and quackery too: "*But as it were to balance this immediate Help of Providence, nothing was otherwife wanting to aggravate the common Deftruction; and to which nothing more contributed than the Practice of Chymifts and Quacks, and of whofe Audacity and Ignorance it is impoffible to be altogether filent; they were indefatigable in fpreading their Antidotes; and although equal Strangers to all Learning as well as Phyfick, they thruft into every Hand fome Trafh or other under the Difguife of a pompous Title. No Country fure ever abounded with fuch wicked Impoftors; for all Events contradicted their Pretenfions, and hardly a Perfon efcaped that trufted to their Delufions: Their Medicines were more fatal than the Plague, and added to the Numbers of the Dead...*" Thank heavens for modern medical regulation!

Elsewhere in *Loimologia* Hodges provides further insight in the activities of some individuals: "*Some old Nurſes, as themſelves have informed me, for an Antidote gave human Excrements; but for the Efficacy of this Secret, I have nothing to ſay. Some found Benefit by drinking of Urine; but many who have thought themſelves by theſe Means ſo well fortified, would venture themſelves too inadvertently into Danger, without any neceſſary Occaſion, to the great Hazard of their Lives.*"

William Boghurst, who was an apothecary and contemporary of Hodges during the Great Plague of London, notes in his *Loimographia* [1666] how the flight of trained physicians out of London during the plague left the door open for quacks to step into the medical frame: "*... ſweating was the moſt generall course taken, yet were not other ways left unattempted by rude and ignorant people, as purging, vomiting, bleeding. For the phyſitians almost all going out of Town left the poor people for a prey to theſe devouring blunt fellows, blind Bayards, who were very bold to turn Doctors. The abſence of thoſe who ſhould have watched over the people's lives handed them the opportunity by their going aſide, and theſe were all ſuch timber-headed fellows that not withſtanding their confident fearleſs familiarity they could make noe accurate obſervations at all, which the Phyſitians could eaſily have done, had they the ſame courage, and the new upſtart Empiricks made ſad work both with people's bodyes and purſes, ſelling their idle medicines at an extraordinary dear rate. One ſold purging pills at half a crown a pill, and an ointment at 5d. an ounce, theſe also ſold pomanders very dear and playſters 5 Shillings a piece. One gave Mathews his pill to ſweat, made with hellebore, opium, etc., which choaked people... One Gray, a quaker in Gutter lane, gave purging Pills to all that came to him; a great many alſo that would not venture in the battle would hold themſelves excused by writing ſingle ſheets of directions out of charity, which did as much hurt as the Mountebanke; for though they never ſaw one ſick yet would undertake to write that vomits was the only way, others that bleeding was good. Some were for purging, but not to regard what they did here or beyond ſeas...*"

Boghurst's comment on the rise of the medical pamphleteers towards the end there, brings to mind a medical self-help book published by the preacher John Wesley, *Primitive Physick*. Although there are possibly some useful remedies in the work, which was first published in 1747 and ran to many updated editions, it has all the hallmarks of someone trying to profit from their position in society – exactly what happens today, where celebrities are asked to pen an article or book about a subject they have little, scant, or no knowledge of.

For asthma Wesley recommends (in the 1750, 2nd edition) drinking a pint of cold water every night as you lay down to bed, and the same was also suggested for treating a cold. Another asthma tip was: "*Or, a Pint of cold Water every Morning, washing the Head therein immediately after, and taking the Cold Bath once a fortnight... or drink Sea Water every morning.*" A cold bath was suggested too for deafness, and he also says blindness '*Is often cured by Cold Bathing.*" One supposes that in less developed parts of the world even to this day, there are equivalents of Wesley peddling nonsense to the unwitting masses in their thrall.

When it came to treating the plague, one of most terrifying of all diseases in past centuries, Wesley simply has this advice for his readers: "Cold Water *alone, drank largely, has cured it... Or, a draught of* Brine *as foon as feized; fweat in Bed; take no other Drink for fome Hours.*" Wesley does not cover plague in a 1750 (2nd) edition of the work, while the extract above comes from the 1762 edition.

Another dubious compendium of medical know-how, trawled from wherever, was *The Treasure of Health* [1819] which circulated around North America at the beginning of the 19th century. The author, Lewis Merlin (perhaps the name should suggest something there) has two extraordinary plague remedies using toads; though admittedly live toads and dried toad have featured in remedies over the centuries, including some to treat plague. Here's what he has to say to his readers:

"*Remedy against the Plague.*
Put into a glazed earthen pot twenty or thirty large loads, cover it with its lid, and apply clay all around it, after tie it on the pot with wire, then put the pot on a fire of coals made in the middle of a large yard, leave it seven hours on the fire, after which you must take it off and let it get cold, afterwards, open the pot; having previously put a handkerchief before the nose, for fear the smoke should get into the head. In the pot you will find a powder, gray and white; both have the same property. Put some of this powder in a small glass of white wine, and on the next day give it to the patient who has the plague; three hours after, he will experience a great heat all over his body, which will last two hours; his clothing must be changed in bed, and as soon as he does not sweat any more, he must take a meat broth.

Another Magnetic Pentacule, or Medal, a preventive for the poor,
    Roast on both sides a piece of bread, as large as the palm of your hand, and half an inch thick, until it is very well dry; afterwards you must pick it on both sides with the point of a knife; then, put it under a toad which you must roast alive to receive its grease, sometime on one side and sometime on the other of the bread, until it is quite well imbibed with it. Afterwards cover the bread with two bits of cloth, and wear it between the dress and shirt on the region of the heart.
    This is the preventive generally used by those who are exposed in taking away and burying the bodies of those afflicted with the plague."

Well before Merlin's work, the author of *Lectures on Materia Medica* [1770], and probably others too, had questioned the benefit of using toads, although earlier medical luminaries such as the Swiss-German Paracelsus, living in the 16th century, believed toads could heal pestilential buboes in the groin, while the Flemish chemist van Helmont, who lived roughly around the same period, had made a plague amulet from toad. Here's what *Lectures* has to say:

"Toads are commended many ways. Bates ſays of his pulvis Æthiopicus, made of toads burnt alive, (horrid!) "Doſis 3ſs. vel ultra, in variolis, &c. vel in moribundis auxilio certe fuit." Oleum Bufonum is ſaid to be good for the dropſy and ſcrophula outwardly; ſp. vini, or wine in which ſome Toads have been drowned, to be an egregium & expertiſſimum alexipharmacum, inwardly; wonderful amulets to be made of the creature itſelf; and what not... All the favour I aſk for theſe innocent (though not very lovely) animals is to kill them before they are burnt; a favour never denied to the created criminals; for I can aſſure you the powder will be nothing the worſe for it. But now poor Bufo is diſmiſſed ſafe."

Looking at woodcuts in old herbals one is apt to forget that the apothecary was in a 'business' that had its own economic dynamics. Boghurst mentions the price of rue and scordium going up in plague times while *Compleat Body of Distilling* [1731] provides readers with production costs for bulk manufacture of medicinal, and other, waters and spirits.

THE PLANTS

Some of the plant profiles contained here are relatively short but are included since their herbal use presents either an unusual, or uncharacteristic, adoption of a particular plant that is well-known today by plant enthusiasts. Other profiles are much longer since that species was a key ingredient in anti-plague remedies.

One observation that the reader may well find themselves concluding upon reading through the following text extracts is how tenuous the connections were between a plant's supposed herbal attributes and the reality. William Buchan, writing in his *Domestic Medicine* [1790], summed it up like this: "*Ignorance and ſuperſtition have attributed extraordinary medical virtues to almoſt every production of nature. That ſuch virtues were often imaginary, time and experience have ſufficiently ſhewn. Phyſicians, however, from a veneration for antiquity, ſtill retain in their liſts of medicine many things which owe their reputation entirely to the ſuperſtition and credulity of our anceſtors.*" Keep that thought in mind as you travel back in pre-germ theory and pre-antibiotic times, and see what was served up to neutralize the deadly killer effects of *Yersinia pestis*.

Regarding some of the recipes with weights and measurements displayed in ounces, scruples, drachms, and such like. Pharmacists and apothecaries of old used Troy weights rather than Avoirdupois; so a pound is 12 ounces under Troy, one ounce eight drachms, one drachm equals three scruples, and the scruple weighs in at twenty grains. The only time when Avoirdupois really came into use in pharmacy was in the matter of bulk purchase of ingredients. Not taken into account here is that different countries and medical systems often attributed different weights to an ounce, scruple, and so on.

For a reminder of old medical terms that appear in the descriptions refer back to the short glossary on page 20, while the following list should help refresh your memory on ingredient measurements in the old remedies that follow:

g, *gr* – grain	*i*, *j* – one
Э – scruple	*s.s*, *s*, *ʃ.s*, *ʃs*, ß, ẞ – half
3 – drachm	Numero, N, Nº – a number of
Ʒ – ounce	Ana, ā, an – of each
lb, *℔*, *lib*, *li* – pound	Partes æquales, part. aeq., p.a – equal parts
m, *M* – handful	Quantum sufficit, q.s. – sufficient quantity
p, *pugil* – fist	Secundum artem, S.a. – the second art
gut, *gt*, *gtt* – drop/s	R, Rx – recipe
coch, *cochl* – spoon	

Key to the layout of the text that follows here is the chronological use of herbal plants during plague, to allow better analysis and comparisons of the data. For this reason the running order is mostly laid out sequentially for each plant species, with the first or prior publication date of a recipe or reference used as the date benchmark, not the re-published date. This only concerns a handful of texts, mainly those of Ambroise Paré, William Bullein, Thomas Willis, Nathaniel Hodges, and a few others. There are prior editions of some of the texts used to compile this book but they are often highly elusive, and those included here are those to which this writer has obtained access. After each of the text extracts the source is given in square brackets in the hope that it will make life easier for readers to scan quickly through the content for their own researches.

For readers unfamiliar with my background in edible wild, stroke survival, food with Wild Food School, this book is a secondary outcome of still ongoing researches into the latter subject, but would not exist without the extraordinary resources of our British Library, its librarians, curators, and the millions of manuscripts and rare books held in the archives of that wonderful resource. Even today, in the age of digital media, there is nothing that quite ticks the box like picking up a scribe-written book five or six hundred years old.

WILD ANGELICA

BARBERRY

BETONY

BLESSED THISTLE

CARLINE THISTLE

BORAGE

SALAD BURNET

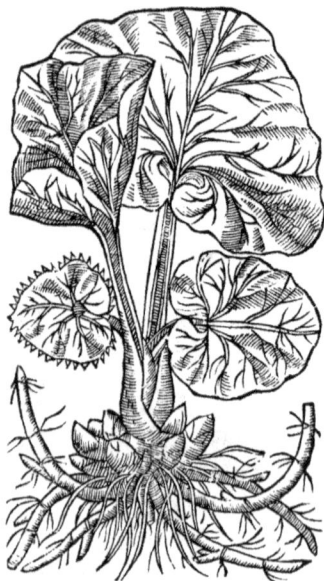

BUTTERBUR

ANGELICA – *Angelica archangelica* and *sylvestris*

Angelica is a plant that features very widely throughout herbal remedies prepared to combat the plague, and it would appear that both the domesticated variety [*Angelica archangelica*] and the wild variety [*A. sylvestris*] were used. In the wild angelica grows in low-lying ground or hollows that are constantly moist and are often shaded, but will be found bordering exposed roadside ditches, wet ground near streams, and marsh borders. Often characterized by claret-coloured stems and flower masses, where exposed to strong sunlight, wild angelica is a herbaceous perennial that will grow to a height of five or six feet under optimum conditions, and generally flowers around June to August. Green [1823] tells us: "*In gardens near London, through which small streams of water run, great quantities of this plant are propagated, the tender stalks of which are cut in May for the confectioners, who have a great demand for it as a sweetmeat.*"

Old botanical accounts about the plant suggest angelica acquired its name because of supposed magical properties including anti-witch propellant qualities. Called the Herb of the Holy Ghost its' angelic-like herbal properties meant that angelica was considered good against poisons, pestilent agues and the pestilence, bites of rabid dogs and subsequent hydrophobia.

The stems of garden variety, but also of the wild type, can be candied in sugar to be used as sweetmeats, in cakes, or as decoration on trifles and cup-cakes. A yellow dye can also be obtained from the plant which no doubt was thought to give cloth stained with it some divine attribute. The 17th century apothecary-botanist John Parkinson says that: "*In Sussex they call the wilde kinde* (of Angelica) *Kex, and the weavers winde their yarne on the dead stalks*". It should be pointed out that 'kex' is a name commonly given to several members of the *Umbellifer* family – of which angelica is one – including poisonous hemlock. So beware potential misinterpretation of old botanical texts.

1527 – An early English translation of Hieronymous Brunschweig's book on distilling (originally published in 1500) provides the following insight into the use of angelica water as a plague remedy: "*The beſte tyme and parte of his dyſtyllacyon / is the rote in the ende of the ſeconde yere in the herueſt / chopped / ſtamped and dyſtylled. A water of Angelica is the mooſt worthyeſt water that may be founde agaynſt the peſtylēce. Yf therof be dronken halfe an ounce eury mornynge faſtynge. And whan any body is taken with the peſtylence / he ſhall take of the ſame water two oūces / Tiriaca geneſti one dragma / powder of the rote of Angelica halfe a dragma / vynegre a quatre of an oūce. Theſe ſhal be mixed eche amōge other / and that ſhall be gyuen to the ſeke body / or euer he ſlepe / but fyrſt he ſhall be lette blode in the place that is moche neceſſarye. And whan he hath dronke that for named drynke / than he ſhal be layd downe / & well couered that he may ſwete / for that is to hym a grete helpe.*" [Vertuous Book of Distillation]

1551 – Regarding angelica root Textor says: "*Pour fouuent tenir en la bouche & mafcher, mefme quand on fort & on côuerfe auec les gens, il y ha efcorce & femêce de citron, qui baille aufsi bonne odeur cinnamome ou canelle, gyrophle, racine dangelica ou dauftruche ou de Zedoar, & de femblables deffus nõmees.*" [Maniere Preserver Pestilence]

1578 – Henry Lyte's English translation of Dodoens, the Flemish herbalist, highlights both the wild and domesticated varieties of angelica being used for plague remedies: "*The late writers fay, that the rootes of* Angelica *are contrarie to all poyfon, the Peftilence, and all naughtie corruption, of euill or infected ayre.*
    If any body be infected with the Peftilence or plague, or els is poyfoned, they giue him ftraightwayes to drinke a Dram of the powder of this roote with wine in the winter, and in the fommer with the diftilled water of Scabiofa, Carduus benedictus, *or Rofewater, then they bring him to bedde, and couer him well untill he haue fwet well.*
    The fame roote being taken fafting in the morning, or but only kept or holden in the mouth, doth keepe and preferue the body from infection of the Peftilence, and from all euyll ayre and poyfon." [New Herbal]

1579 – William Bullein, in his layman's medical work (done as a 'dialogue' between Marcellus and Hilarius), again refers to domestic and wild angelica, and elaborates on the dosages: "*Thys is called the Angels herbe, whych is of ij. kyndes, of the garden and of the field: this herbe excelleth al* [sic] *other agaynft poyfon, and is hoat and dry. It doth open, warme, diffolue, & is good agaynft the fearefull daungerous plague, called the peftilence, if it be but bitten upon, mutch more it is effectual being drunke in the morning. The pouder drunke wyth Wyne. x. grane waight, or the water drunke. xx. droppes in the mornynge in Wyne, is a goodly armour agaynft poyfon, foule ayre or plague...*" [Bulwarke of Defence]

1585 – Continental Europe, at the same period, also had angelica in its sights for its anti-plague properties. This from Castor Durante's *New Herbal* published in Rome: "*... ò d'altri mali contagiofi con acqua di cardo fanto, di tormentilla con vn poco d'aceto & con vn poco di Teriaca. L'Acqva deftillata dalla radici vale à tutte le cofe predette: conferifce mirabilmente nella contagion delle pefte, & nelle febri peftifere, & furiofe pigliandofene meza dramma con vna dramma di Teriaca in quattro once de queft' acque facendo poi fudar l'infermo: vna dramma delle fua poluere puo fupplir per Teriaca.*" [Herbari Nuovo]

1596 – In a less widely known English health text, but again written in a language that made the knowledge accessible to non-medical readers, there is mention of: "*A marvellous good drinke for them that are infected with the Plague..*
Take leafe-gold, and mingle it with the iuice of Lemons, and a litle Suger-candie, Cloues, Mace, and a litle Cinamon, and like quantity of Licquorice finely pared & fliced, and let this be fteeped in white Wine, or elfe in good Claret Wine, and put

therein a good quantity of the powder of Angellica [sic], *or elſe of the decoction of the ſame roote, the ſame drinke will help the Patient being drunke warme."* [Rich Store-House or Treasury for the Diseased]

1617 – Woodall suggested that ship's surgeons of the East-India Company should have angelica water in their medicine chest, but the most interesting insight are his comments on the potential shelf life of the made product: *"Angelica water may ſerue well in ſtead of* Trekell *or* Mithridate, *for a preſeruatiue againſt the plague or any infectious aires, for there is no one thing more commended by ancient and moderne writers, in that kinde, then* [sic] *Angelica is... and being truly made will retaine his ſtrength and vertues forty yeeres and more."* [The Surgions Mate]

1625 – Thomas Thayre's short layman's work entitled *An Excellent and best Approved Treatise of the Plague* was published in London during the same year that London was visited by a significant plague outbreak that killed an estimated 35,000 people. Thayre was a medical practitioner and his book (which has a Galenic approach to plague as a contagious illness), is aimed at the literate among the general population and is typical of many publications that appeared in such periods of history. The advice that Thayre provided may well have been used on a daily, needs-be, basis by London's inhabitants. Among a number of general preservative remedies against infection the author has:

"Another very good Preſeruatiue, and worthy of much commendations.
R. Of good Mithridatum halfe an ounce, Angelica root in powder two drammes, of Theriaca andro. halfe an ounce, Bole-armoniack praep. two drammes, conſerue of Roſes and Borage halfe an ounce, ſeede of Citrons two ſcruples, ſirup of Lemmons one ounce, mixe them, make halfe this receipt." Interestingly, the use of borage in anti-plague remedies is pretty scant, though it does appear among the recommended dietary greens during pestilential times.

Elsewhere, Thayre has a remedy that he describes as a 'princely preservative' which contained many ingredients and would have been potentially expensive, while providing others for 'commons and countrymen'; a clear indication – if any were really needed – that the poor had little chance of medical care if they could not afford it. Indeed, the previous recipe with lemon syrup must have been some-what costly at the time. Angelica, meanwhile, regularly appears in many of Thayre's home-help remedies, and among those is one listed as a 'good preservative' that also features syrup of lemons and which he prefaces with: *"Another that defendeth all men that vſe it, from the infection of this contagious ſickneſſe,"*

R. Theriaca Andromachi, mithridatum optimum ana two Dramms, confer. Roſarum three drammes, Boli armeni praep. two ſcruples, ſem. vel rad. Angelica two ſcruples, ſem. citri halfe a dramme, ſir. Limonum halfe an ounce, miſce.

Take of this euery morning, the quantitie of a hafel nut, or any other time of the day, if you goe among any throng of people, or where the fickneffe is, but you ought to faft after it a while."

An example of the division between the treatment of rich and poor can be seen in another of Thayre's remedies, this time in a 'sweat' for the 'Commons': "*R. Mithridatum two drammes, Venice treakle one dramme, mixe them with water of Angelica, Cardus Benedictus or Scabious, or for want thereof poffet drinke, made with white wine, and fweat well.*" [Excellent Treatise of the Plague]

1665 – Move the clock forward to the next period in English history where a serious plague outbreak affected the population, the Great Plague of London. While there are contemporaneous accounts of the medical events in London from the hands of physicians such as Nathaniel Hodges, the following extract comes from Robert Lovell's herbal work *Panbotanologia* which was published in Oxford, and where there was a considerable body of medical knowledge and learning at the time. Lovell seems to prefer garden angelica over the qualities provided by the wild variety, and refers to a 'water' from the London Pharmacopoeia as follows: "*Spiritus & Aqua angelica magis compofitae pharm: Lond.* or *Spirit,* and *water of Angelica of the greater compofition,* ftrengthens the heart, and refifts infections; It's good in time of peftilence and ftinking aire. That in the *former difpenfatory,* comforts the heart, cherifheth the vitall fpirits refifts the peftilence, and corrupt aires, which caufe epidemicall difeafes: the Dose is to the fick *Cochl.* i. in a convenient cordial; others that are cold may take it fafting or before meat."

Lovell highlights angelica root as being useful to chew in the mouth and that through its diuretic and sweating properties it helped in pestilent fevers, specifying a dose of one drachm of powdered root drunk in 'thin wine', or in the distilled water of carduus benedictus or tormentil, or vinegar or treacle.

1666 – Published in 1691 the recipes for plague and fever cures in *A Plain and Easie Method* are interesting in that they represent remedies used in practice by Thomas Willis during the 1665-1666 plague outbreak. The text has angelica root used in the following two-stage remedy: "*For fome that are of a hotter Constitution, and a high Sanguine Temper, it may be proper to take every Morning a fpoonful of Peftilential Vinegar in a little Carduus water, or plain Wallnut water; or elfe drink a draught of Poffet-Drink, made with a Spoonful of that Vinegar.*

Let the *Pestilential Vinegar* be made thus.
Take of the Roots of Angelica, Butter-Burr, Tormentil, Elecampane, of each half an Ounce, Virginian Snake-Weed, choice Zedoary, Contrayerva, of each three Drams, Leaves of Scordium, Rue, Goats-Rue, of each one handful; Marigold Flowers, Clove-gilloflowers, of each half a handful; Seeds of Citron and Carduus, of each two Drams, Cut and bruife thefe, and put them in a Glafs-Bottle, with three Pints of the beft Vinegar, to digeft for ten days." [Plain & Easie Method]

1713 – From the next century is a recipe in Bate's *Pharmacopoeia* to make a medicinal amulet against the plague. The original work by George Bate (1608-1669) was translated from Latin into English and subsequently reproduced by William Salmon in 1713, although from the introductory texts in the reprint, it appears Salmon had been working on his version in the early to late 1690s. Salmon adds his own notes below each of Bate's remedies, sometimes adding his own alternative recipes; however as a medical practitioner, Bate lived through the period of London's Great Plague. In this following instance Salmon does not appear overly convinced regarding 'amulets' against plague, while *auripigment* is orpiment, or yellow arsenic:

"*Amuleta Peſtilentiale,* A Peſtilential Amulet or Preſervative.
Bate.] Rx *Auripigment* ʒiſs. *Angelica Roots* 3 vj. Mucilage *of Gum Tragacanth, q. s. mix and make a Paſte, of which make Amulets, Nº xij. of a Globe like Form, which roul up in Silk,* and hang againſt the Region of the Heart.

Salmon.] *Paracelſus, Crollius, Hartman,* and others, are of the *Opinion,* that there is a mighty Power, Force, and Virtue, in ſuch like Amulets, made to defend againſt the Infection of the Plague; I will not ſay much to the Matter, but leave every one to uſe their own Freedom...
However this I will ſay that if my own life lay at ſtake, I ſhould rather truſt to ſome known and approved inward Specifick Antidote, than to all the fam'd Amulets in the World." [Pharmacopoeia Bateana]

1721 – The French city of Marseilles was hit by a plague epidemic in 1720 and a number of medical works on the subject appeared soon after; with English physicians of the early 18th century eager to learn from the medical experiences of their counterparts across the Channel. *Avis de Precaution contre la Maladie Contagieuse de Marseilles*, sometimes simply referred to as '*Avis de Precaution Contre la Peste*' appeared in various formats at the time, French editions being printed in both Lyon and Turin. Given the close proximity of Marseilles to the Mediterranean region the citrus fruits used in the following treacle water recipe would have been cheap when compared to the same 'water' made up by English physicians and apothecaries, while *oseille*, or sorrel, was a commonly used culinary vegetable in French cuisine:

"*Eau thériacle... Prenez des racines d'angélique, de zédoaire, de dictam blanc, de meüm, de chacune une once, des feüilles de veronique, de charbon* [sic] *bénit, de ſcordium, de ruë, des ſommitez de mille pertuis, de chacune une poignée, de ſemences de citron, d'oſeille, de coriandre, de chacune deux dragmes, du canfre, du ſafran, de la mirrhe, du macis, de chacune une dragme, un citron coupé par tranches, de la thériaque, quatre onces.*

  Mettez toutes choſes préparées dans un vaſe de grandeur ſuffiſante, arroſez le tout d'eſprit de vin tartariſé, & laiſſezle en digeſtion pendant deux jours. Ajoûtez quatre pintes de vin blanc, & diſtillezle au bain-marie." [Avis de Precaution]

1747 – When James' extensive pharmacopoeia was published in the mid-18th century it would appear that the herbal emphasis on angelica as a key anti-pestilential remedy ingredient was in decline. In the following passage there is a passing reference made to angelica roots for a plague 'sweat', but the focus is far more on angelica's other herbal uses:

"*Angelica sativa: This is a Plant of an highly penetrating and aromatic Nature; its Seeds and Roots are in a particular Manner refolvent and ftimulating, and confequently fudorific, alexipharmic, and proper to expel the peftilental [sic] Poifon by Sweat. The Root is thought beft, which when chewed, has the Tafte and Smell of Ambergreafe and Mufk mixed together, and fpreads a Kind of penetrating Gratefulnefs all over the Mouth, without exciting any Inflammation. Hence an Infufion or a gentle Decoction of it, is commended againft a fetid Breath, and when ufed in the fame Manner, it is faid to be beneficial in Coughs arifing from Cold, or a vifcid Mucus; becaufe it renders Refpiration more free and eafy. From what has been faid we may know, why the whole Plant is claffed among the carminative Medicines, and for what Reafon fome recommend a Dram of its dried Powder, taken with Wine or Rob of Elder, in intermittent Fevers. In Medicine the Root is more frequently ufed than the Seeds, whilft the Leaves are entirely neglected.*" [Pharmacopoeia Universalis]

1751 – John Hill, self-styling himself Sir John after he was awarded a Swedish scientific prize for contributions to botany, wrote and published a number of populist herbal medicine books during the mid-18th century. In his *History of the Materia Medica*, obviously aimed at a herbal-medical readership, Hill casts aspersions on the perceived sudorific qualities of angelica thus: "*Angelica is ftomachic, cordial, and fudorific, and is greatly recommended againft peftilential Difeafes, and the Plague itfelf, but the Virtues afcribed to it on this Account are fomewhat too great. It has been made an Ingredient in many of our officinal Compofitions; but the late Difpenfatory of London, has omitted it in many of them: They have however retained the Leaves of the Plant in the* Aqua Alexiteria Simplex & Spirituofa, *and in that* cum Aceto, *which is intended to ferve in the Place of the Treacle Water of the former Difpenfatories. The Stalks make a very pleafant Sweet-meat preferved with Sugar, and this is a very good Way of taking Angelica on many Occafions.*" [History of the Materia Medica]

Oddly, the text of Hill's first anonymously authored herbal for general public consumption three years later – *The Useful Family Herbal* [1754] – is suggestive that angelica is still useful in plague remedies, or at least that is the impression this writer has in reading through Hill's text: "*Every Part of the Plant is fragrant when bruifed, and every Part of it is ufed in Medicine. The Root is long and large; we ufe that of our own Growth frefh, but the fine fragrant dried Roots are brought from* Spain. *The whole Plant poffeffes the fame Virtues, and is cordial and fudorific; it has always been famous againft peftilential and contagious Difeafes. The Root, the Stalks candied, the Seeds bruifed, or the Water diftilled from the Leaves, may be ufed,*

but the Seeds are the moſt powerful. It is alſo an Ingredient in many Compoſitions."
[Useful Family Herbal]

1761 – Clergyman John Wesley also weighed in on the medical world, publishing various editions of a book entitled *Primitive Physick: or, an Easy and Natural Method of Curing Most Diseases.* Some of the remedies in this work are highly dubious but in a cure for the plague he says: "*... take a dram of* Angelica *powder'd every ſix Hours. It is a ſtrong Sweat."* [Primitive Physick]

1770 – By the time of the publication of Alston's *Lectures on the Materia Medica* from around the same period, mention of angelica in cases of the plague has largely vanished. Instead, most references are to its' general diaphoretic-sudorific qualities: "*Angelica is an acrid aromatic attenuant, diaphoretic, cordial, carminative, antiſeptic, cephalic and alexipharmic; and commended internally in female diſeaſes, menſibus lochiiſque obſructis, partu difficili, ſuffocatione uteri, poiſons, venomous bites, malignant diſeaſes, the peſtilence not excepted, &c. and externally as diſcutient and anodyne, for inflammations, gangrene, tooth-ache, &c.*

The taſte is very hot, diffuſing a glowing warmth through all the mouth, the lips themſelves being ſenſibly affected with it: yet it is ſweet and agreeable, the very bitterneſs improving as it were the flavour. The pungency of it will be felt more or leſs for ſome hours: the bitterneſs not many minutes. Some think it ſmells of muſk."
[Lectures on the Materia Medica]

1790 – Again, in William Meyrick's family herbal, the reference to angelica in plague seems to be put on a par with its use in treating ague and other complaints, though it must be remembered that by 1790 the public's memory of serious outbreaks of plague upon British shores were largely a thing of the past: "*Every part of the plant is full of virtue, but the roots and ſeeds poſſeſs the higheſt degree. They are cordial, ſudorific, and ſtomachic, of great efficacy in peſtilential and contagious diſorders. They are likewiſe ſerviceable in all cold flatulent complaints, and ſeldom fail of removing the ague, if taken three or four times repeatedly on the approach of the fit.*

A ſcruple of the dried root in powder, or ten grains of the ſeed is a moderate doſe. The roots and ſtalks are ſometimes candied, and in that ſtate they are equally efficacious, and more agreeable to be taken." [New Family Herbal]

1822 – One of the most interesting early 19th century works on the plague is written by a Frenchman, Martin de Saint-Genis, who takes a highly analytical approach to the medical accounts of past plague outbreaks, comparing recorded symptoms over the ages, and so on. Despite this his *Manuel Préservatif et Curatif de la Peste* still hearkens back to the past in places and we find him recommending a number of Diemerbroek's old potions from the 17th century. Among the plague-remedies with angelica are a vinegar and preservative wine, but also mention of chewing angelica root, plus some rather scathing comments on anti-plague amulets as follows:

"*Vinaigre impérial. PRENEZ un pot de vinaigre fort, (le blanc est le meilleur), racines d'angélique, d'impératoire et clous de girofle légèrement concassés, de chacun deux drachmes; mettez le tout ensemble dans une bouteille de verre bien bouchée, et après l'avoir bien agitée, pour mieux faire le mêlangé des drogues, laisser cette bouteille sur les cendres chaudes pendant une nuit et la conserver. Le matin avant de sortir on s'en frottera la face, les mains et les poignets; et pendant la journée on en mettra dans un flacon pour le flairer continuellement.*

Vin préservatif
METTEZ dans de l'excellent vin vieux des racines d'angélique [,] de zédoaire [,] d'aunée, dès sommités de melisse, des fleurs de scabieuse, d'oranges, de roses rouges, d'écorce de citron, dès baies de genièvre, de la canelle, du safran, des clous de girofle, laisser le tout en digestion pendant plusieurs jours, ensuite couler et passer à travers la manche. Pour vin à prendre le màtin à jeun, à la dose d'une ou deux cuillerées à bouche.

Il est utile de mâcher ou rouler très-souvent dans la bouche, des morceaux de racine d'angélique, d'impératoire, d'aunée, ou des baies de genièvre, ou des clous de girofle, ou de l'écorce sèche de citron et d'orange, ou de mâcher des feuilles de tabac desséchées. Il faut avoir l'attention de ne point avaler sa salive, lorsqu'on est dans la chambre d'un malade.

Il faut se défier des charlatans et des prétendus secrets qui préservent ou qui guérissent de la peste; ainsi que toutes les amulettes que la crédulité, la super-stition, et la peur accréditent. Les magistrats doivent soigneusement éloigner tous les charlatans, qui ajoutent à la malignité de l'épidémie en communiquant avec une multitude d'individus sains ou malades et sans aucune précaution, et vendant des drogues nuisibles." [Manuel Préservatif et Curatif de la Peste]

1852 – By the time of the 11th edition reprint of Thompson's *London Dispensatory* in the mid-19th century angelica clearly holds no sway among official medicines, although the work does outline the medical properties and uses of angelica thus: "*The leaves and fruit, or seeds, when recent, and the root both in the fresh and dried state, are tonic and carminative; but although the most elegant aromatic of northern growth, yet they are scarcely ever prescribed in modern practice. The dose in substance is from 3ss. to 3j., which may be given three or four times a day.*"

And as part of the general introduction into the use of angelica for medicinal purposes the text says: "*... the roots should be dug up in the autumn of the first year, as they are less liable to spoil than when they are taken up in the spring; for when gathered at that season, they become mouldy, and are preyed on by insects. They should be thoroughly dried, and kept in a well-aired, dry place; and in order to secure their preservation Lewis suggests "the dipping them in boiling spirit, or exposing them to steam, after they are dried." The leaves and fruit do not retain their virtues when kept.*" [London Dispensatory, 11th Ed.]

BARBERRY – *Berberis vulgaris*

Sadly it is hard to find the true, native, barberry shrub growing in the wilds of Britain, although there are ornamental varieties often to be found in tame garden environments. The reason for the demise of barberry in the wild is that from roughly the middle of the 18th century onwards the shrub was associated with harbouring wheat rust that compromised wheat yields. At a period when agriculture was starting to be approached more scientifically this rust-host connection resulted in barberry bushes being grubbed out of countless country hedgerows and across Britain's farmland.

Despite its rarity in farmland barberry is dispersed widely across the British Isles so might be found in pockets of woodland and coppices. Preferring a humus soil, but tolerating sandy soil too, this perennial shrub grows best in a peaty loam or sandy-humus, and reaches a height of eight to ten feet, flowering around April to May. The juicy, acid-tasting, red berries can be made into conserves, and were once used in the kitchen to produce a tart sauce to accompany meat, as a garnish for roasts, and also to make iced sorbets. Green, in *The Universal Herbal* [1823], tells us: *"The Barberry is cultivated for the sake of the fruit, which are pickled, and are used for garnishing dishes, and being boiled with sugar, form a most agreeable rob or jelly; they are used likewise as a sweetmeat, and in sugar-plums or comfits."*

Barberry had a variety of common names ranging from Barbaryn, Berber, Jaundice Berry, Maiden Barberry, Pepper-ridge, Piprage, Woodsour, Woodsore, Woodsower Tree, some of which are similar to completely unrelated plant species. In *Modern Husbandman* [1750], Ellis accounts for the name Jaundice Berry as follows: *"The wood of this tree is said to be such an antidote against the Yellow Jaundice that, if a person constantly feeds himself with a spoon made of it, it will prevent and cure this disease while it is in its infancy."* And under the *Doctrine of Signatures* the yellow bark was also considered as a cure for jaundice, being purgative and taken in ale for this purpose, while the astringent scarlet berries were eaten to remedy bilious and stomachic disorders.

1551 – In Textor's work on the plague barberry appears in a snippet with regards to allaying thirst in plague fevers. The name for barberry in French is *vinette*, but Textor also calls it *oxycantha*, which these days we would more readily associate with hawthorn:

"... Au lieu dun tel vin il y ha celuy defpine vinette dite oxyacātha, vulgairement berberis. Ou prenez ius doranges, & faites cōme iay dit des grenades. Ou prenez des autres ius, comme doxeille, ou daigret, ou des deux & defemblables. Ainfi compofez potions de diuers ius car ilz font bien vtiles." [Maniere de Preserver de la Pestilence]

Interestingly, the first published English translation of the *Kruydeboeck* [1554] – originally written in Flemish by physician Rembert Dodoens – was produced by

Henry Lyte in 1578 and makes no mention of barberry for 'plague fevers'; the text merely refers to barberry as a general fever simple: "*With the greene leaues of the Barberie buſh they make ſawce to eate with meates as they do with Sorrel, the which doth refreſh and prouoke appetite, and is good for hoate people and them that are vexed with burning agues.*" The translation also tells us that in the home-ground of Dodoens, the Brabant region of modern day Belgium, barberry was: '*... muche planted in gardens, eſpecially in the gardens of Herboriſtes.*' Somehow it seems at variance that the Flemings were not making more use of barberry in plague, and yet barberry was being cultivated as a herbal ingredient. It should be pointed out that although we think today of Belgium as being one national State, in the past it was a patchwork of independent French and Dutch speaking principalities, so there may not always have been interconnections between the local scientific and medical communities.

1583 – A further French text of a similar period – *Excellent Traicte de la Peste* – has barberry as one of the acid accents used in diets during time of plague, rather putting the plant into the framework of medicinal cuisine, a topic discussed in an earlier section: "*La ſaulce dicelles ſera verius, vinaigre, ius de limons, orenges, citrons, grenades aigres, eſpine vinette, groſeilles rouges, & vertes, ius d'ozeille champeſtre, & domeſtique.*"

In a section entitled '*Deſcription des eaux cordiales preſeruatiues, & Curatiues*' we also have the following composite plague-water recipe:

"Rx. Aquæ vitæ ℥ xij. Succi berberis ℥ vj. Succi calendulae ℥ viij. Theriacæ Andromachi ℥ iiij. Radicis Gentianæ, Angelicę, Tormentillæ, Corticum citri, Ruthæ an. ℥ vj. Boli armeni opt. ℥ ſ.
Simul macerentur per dies duos in loco calido: tum diſtillentur in cineribus, igne lento. Doſis ℥ j. in aqua conueniente." [Excellent Traicte]

1602 – Further mention of *vinette* as medicinal cuisine appears in Caesar Morin's plague treatise where he suggests that the diet should, roughly translated, include: '*... a chicken wing or a capon with verjuice, lemon or vinette.*'

Elsewhere Morin suggests the following for the patient: "*Et en cas que aux iours defendus le patient ne voudroit máger de la chair, il faudra luy donner des oeufs pochés en l'eau, auec du ius de limon ou de vinette.*" [Traicte de la Peste]

1706 – An early 18th century French language reproduction of the works of the 13th century monk Albertus Magnus refers to barberry water mixed with treacle as a plague treatment, apparently being a tried and tested remedy: "*Pour guerir de la peſte, on prendra demi once d'eau de Vinette, un dragme de Theriaque que l'on faira boire à celuy qui ſera atteint de ce mal, on aura ſoin que cette mixtion ſoit tiede, enſuite on couvrira bien le malade, & on le faira ſuer, il eſt certain que s'il n'y a pas long temps qu'il ait la peſte, il en guerira; c'eſt un ſecret approuvé de plusiers bons & graves Auteurs tant anciens que modernes.*" [Les Admirables Secrets d'Albert le Grand]

1710 – There appears to be less of an emphasis on Barberry used specifically for plague fevers in English texts, although it is present as a general fever simple. William Salmon, in his *Family Dictionary*, highlights a rob that can be made from the red berries: "*There is a Rob made of Barberries in this manner: put to the Juice of Barberries one pound, or a pint, half a pound of D.L.S.* [ED - Double Refined Loaf-Sugar] *and with the gentle heat of a Bath make it into a due thicknefs... This is excellent good in hot Difeafes, quenches Thirft, cools the Stomach, and creates an Appetite.*" [Family Dictionary]

1770 – Alston's *Lectures on Materia Medica* has the following: "*It is an acid, antifeptic and fubaftringent, like acetofae folia; called ftomachic by fome, cordial by others; and is commended in hot, bilious, and inflammatory difeafes, as ardent and peftilential fevers, vomitings, fluxes, hemorrhages, &c. The inner bark is an attenuating diuretic and cathartic deobftruent; and commended in obftructions in the vifcera, efpecially the jaundice.*" [Lectures on the Materia Medica]

1790 – Meyrick makes reference to the Italian physician-botanist Prosper Alpinus (1553-1617) when he relates the use of barberry thus:

"*In ardent, and peftilential fevers, and fluxes, the Egyptians make ufe of a diluted juice of the berries, with the moft happy fuccefs. It is prepared by infufing them for the fpace of twenty-four hours in ten or twelve times their weight of water, and then draining of the liquor and fweetening it with fugar, or fyrup of citrons.*" [New Family Herbal]

1822 – Similarly, Waller, in his layman's domestic herbal, makes reference to the Prosper Alpinus source, but is slightly more elaborate on the uses of barberry: "*The bark and fruit of this plant are principally in use in medicine; the former is extolled in diarrhoeas and dysenteries; the latter, on account of their grateful acid juice, furnish a very pleasant and serviceable beverage in fevers, bilious disorders, and scurvy. This fruit is variously prepared; it may be made into comfits, syrup, jelly, or jam. These different preparations may be employed in forming drinks, which in all kinds of inflammatory diseases, scalding of urine, and especially typhus fevers, are taken with the greatest advantage. The Egyptian physicians make great use of it in this last-mentioned disease, and trust the cure almost entirely to it. Prosper Alpinus attributes his recovery from the plague to following the advice of these men, who gave him no other medicine than the syrup of Barberries with the addition of a small quantity of fennel seed added to it... The syrup is made by boiling together a pint of the expressed and cleansed juice of the berries with a pound and a half of fine sugar in a glazed earthen vessel.*" [New British Domestic Herbal]

BETONY – *Betonica officinalis*

Betony, sometimes called Wood Betony, had a long history of use in herbal medicine, particularly for cephalic-related conditions such as head colds and headaches, and was almost seen as virtually a medicinal panacea. Over the centuries, however, betony appears to have fallen out of use because of potential side-effects and violent purging qualities, and just occasionally betony features in plague remedies.

Variously called Betayne, Bidney, Bishopswort, and Wild Hop, betony requires a humus soil and likes shade, so is often found in woodland. However, the plant may also find a home on exposed roadsides, heaths and commons where humus soils exist.

For herbal purposes betony was also used to cure consumption and lung disease, and freshly dried leaves were used as a *sternutory*, to cause sneezing. The leaves have a bitter taste, while the roots are bitter and nauseous, and cause violent vomiting and purging. This, as mentioned, seems to have been a key factor in its demise as a herbal remedy ingredient. As with many plants, in more superstitious times, it was supposed to avert witchcraft, while on a more practical level a dark yellow wool dye was obtained from Betony.

1579 – For all its supposed, miraculous, herbal qualities betony does not really stand out as a key historical plague remedy player, certainly when compared to other vegetable herbal ingredients covered here, such as angelica, butterbur and tormentil. In *Hospital for the Diseased* there is one recipe for a plague water containing betony. This item, incidentally, follows on directly from a recipe for a plague preserving drink containing elder and brier leaves:

"*For the party infected with the Plague*
If it doo fortune one to be fick of ye Plague before he haue drunk the forfayd medicine: then take the water of Scabious a fpoonful, water of Betony like much, & a quantitie of fine Treacle, put it togeather [sic] *and drinke it, and it will expell all the venome.*"
[Hospital for the Diseased]

1593 – In the course of a plague attack there are accounts of patients becoming delerious and raging, and in the following remedy for '*A Sacculus for rauing and raging*' Kellwaye offers a mixture where betony is presumably being included for its cephalic qualities:

"*Florum Nenupharij*, P. j. *Cort: Pap*, 3. ij. *Santali, albi, Rub, Citri, ana.* 3. j. *Florum rof. rub*, P. j. *Florum Viol.* P. ß. *Florum chamomil, Betonicæ, ana.* 3. j.
Shred them all fmall, then pounde then grofly, and quilt them in a bagge, and applie it to the head, and it will helpe you." [Defensative against the Plague]

1665 – Lovell's *Panbotanologia* has very little mention of the benefits of betony in plague although its uses in many other ailments are listed. In the one key plague specific reference where the plant is used, betony features in an ointment for plague sores: "*The juyce inftilled helpeth the pain of the eares,* Park. *with axungia it helpeth plague fores, drunk with pennyroyall and mede it helpeth agues.*" [Panbotanologia, or Compleat Herbal]

1671 – In *Recueil des Remedes et Secrets*, that draws its inspiration from English medical practice of the same period, the following *preservative* against plague appears. This author is unclear whether *Rubus idens* is a misprint and should be *R. idaeus*, or raspberry. *Sureau* is elder, and there is also wood sorrel among the ingredients: "*Prenez Rofes, Betoine, Romarin, de chacun deux poignées: Scabieufe, Eftragon, Sauge, Rhuë, Aceta Ofella, feüille de Rubus idens* [sic], *feüille de Sureau, de chaun une poignée: Bol Armenic trois onces, Safran une dragme, Santal jaune une once, fucre Candy deux onces, tout en poudre fubtile: Diftillez le tout, prenez de cette eau trois cueillerées, & y faites diffoudre la groffeur d'une petite féve de Theriaque du Venife, & le beuvés le matin à jeun.*" [Recueil des Remedes et Secrets]

1720 – Joseph Browne's *Practical Treatise of the Plague* talks about some of the problems with traditional purge remedies and purging methods for some patients, particularly those that could not tolerate aloes well. The following was an aloe-free, betony containing, alternative that he recommended: "*For thofe who cannot take any Aloetick Medicines, or Pills of any Kind, the following Forms are prefcribed. Take Lemon-Peel 3i, Succory, Fennil and Liquorice Roots, of each 3ij, Polipody of the Oak, ʒfs, Annifeed 3ifs, Paeony 3s, Sorrel-Seed 3fs, Endive, Betony, Agrimony and Sorrel, of each a Handful; Carduus, Fumitory, of each half, of the Cordial Flowers, of each a pugil, Sena ʒi; Raifons ftoned, half, Salt of Tartar 3i; boil thefe in two Quarts of Succory Water, to a Confumption of a third Part, and then infufe Rhubarb ℈iv. Agarick 3i, Mace and Cinamon, of each ℈i; ftrain off next Morning, and keep for Ufe.*" [Practical Treatise of the Plague]

1770 – By the time of the publication of Alston's *Lectures on the Materia Medica* betony appears to have no relevance whatsoever to plague cures, if one reads between the lines: "*It is a detergent, deobftruent, and diuretic, and probably alfo cathartic; called cephalic, pectoral, hepatic, uterine: and is recommended internally in the head-ach, vertigo, palfy, the confumption, gout and gravel; and externally as an errhine, or rather ptarmic.*" The conserve of betony flowers was used in head-aches but even this appears to have gone out of fashion by the mid-18th century, the Alston text mentioning: "*Conferva florum betonicæ, & emplaftrum de betonica, were left out of our difpenfatory only in 1744.*" [Lectures on the Materia Medica]

THISTLE, BLESSED – *Carduus Benedictus* / **CARLINE** – *Carlina vulgaris*

Two thistle species rank among herbal plants used in plague remedies, the Holy or Blessed Thistle (*Carduus Benedictus*), and the Carline Thistle (*Carlina vulgaris*).

The first of the two, *Carduus benedictus,* or the Blessed Thistle, is an annual Mediterranean plant species that reaches a maximum of about two feet in height, but is often very much smaller. Like other thistles it has spiny leaves, often being downy earlier in the growing season, and has yellow flower heads. It is a species that prefers broken or cultivated ground, the sands of vineyards, roadsides and waste ground. Older botanical works yo-yo between their scientific naming of the species, sometimes referring to it as *Cnicus benedictus* or *Centaurea benedictus.*

The second species, Carline Thistle, is a plant of dry fields and grassland, sandy heath-like habitats, and even dunes, having yellow-brown to dull red flowers that appear from around June through September. Unlike most of the purple-flowered thistles that we are familiar with in northern temperate Europe the biennial Carline Thistle is generally a short plant, reaching about six to twelve inches in height. By comparison the Cotton Thistle (*Onopordum acanthium*) generally reaches a couple of metres in height – though I once had a specimen in my garden that reached just over nine feet tall. Like *Onopordum* the stems and base leaves of the Carline thistle are downy.

In old times botanical plantlore recounted, rightly or wrongly, that the name 'carline' commemorated the Frankish Emperor Charlemagne who, according to Tabernæmontanus, is supposed to have been visited in his dreams by an angel who fired an arrow from a cross-bow and instructed Charlemagne to use whatever plant the arrow fell upon to treat a pestilential attack running through the ranks of his army. By all accounts the arrow fell on the smaller *Carlina acaulis,* or Stemless Carline Thistle, not *C. vulgaris.* An attribute that both Carline species possess is that the flower bracts lie flat when dry but curl inwards in a moist atmosphere, and so in country districts the thistle was sometimes hung up in houses to serve as a natural weather-glass.

1527 – Brunschweig had a *Carduus benedictus water* in his work on distillation saying: "*The befte parte and tyme of his dyftyllacyon is / the leaue chopped and dyftylled in the ende of the Maye...*" while he instructs that: "*The fame water / preferueth a man of the peftylence / whan therof he drynketh and ounce and a half or two ounces.*" [Vertuous Book of Distillation]

1570 – Charles Estienne's *Maison Rustique* is significant in that this domestic oriented work reflects what French country households and farmers would have been aspiring to in terms of grown or raised produce and food, and general country husbandry. Apart from crops and running a kitchen garden the book also deals with setting up a herb garden of medicinal plants; among them is *Chardon benist*, blessed thistle, and the book provides its readers with notes on the virtues of each herb species to be planted: "*Chardon beniſt demáde ſembable culture que l'angelique… Chardon beniſt n'a moindre vertu contre la peſte, & toute ſorte de poiſon que l'angelique, ſoit pris par dedans ou appliqué par dehors. Ceux qui on la fieure quatre ou autres fieures deſquelles l'acces ſe commence par froid, ils ſont guairis en prenant au matin trois onces de l'eau de chardon beniſt, ou de ſa decoction, ou le poids d'vn eſcu de ſa ſemance pulueriſe.*" [Maison Rustique]

1578 – The Dodoens-Lyte herbal says of carline thistle root: "*The ſame made into powder and taken to the quantitie of a Dramme, is of ſinguler vertue againſt the Peſtilence, for as we may reade, al* [sic] *the hoaſt of the Emperor Charlemaigne, was by the helpe of this roote preſerued from the Peſtilence.*"

The instructions for blessed thistle use were that a: "*Nut ſhell full of the powder of* Carduus benedictus, *is given with great profite againſt the peſtilence: ſo that ſuche as be infected with the ſayde diſeaſe, do receiue of the powder, as is aboueſayde, within the ſpace of xxiiii. houres, and afterward ſweate, they ſhall be deliuered incontinent. The like vertue hath the wine of the decoction of the ſame herbe, dronken within xxiiii. houres after taking of the ſayde ſickneſſe.*" [New Herbal]

1579 – In the dialogue between Marcellus and Hilarius in William Bullein's *Bulwarke of Defence* the Charlemagne tale still persists, but there is dosage detail too for carline thistle: "*… when Charles the great, his armie began to be diminiſhed, through an horrible Peſtylence, he was warned to uſe none other medicine, but only that euery ſouldiour ſhould drinke .3.i. of it in wyne: which beyng done, the ſayd Peſtilence dyd ceaſe preſently.*" Incidentally, Bullein calls the plant the '*Boxe or carle Thiſtell*'. [Bulwarke of Defence]

1617 – Woodall in his *Surgions Mate*, which was written for ship's surgeons on the East-India Company's vessels, recommended blessed thistle to be carried in the surgeon's medical chest, advising that it is: "*… very bitter, and hot, comforteth the heart and vitall parts, moueth ſweate, refiſteth poyſon, is of much vſe in peſtilent diſeaſes, mittigateth the paine of the reines, and ſides, killeth the wormes of the belly, and preuaileth againſt bitings of Serpents.*" [The Surgions Mate]

1629 – The blessed thistle was one of a small number of herbal plants that apothecary John Parkinson recommended in his gardening book for people living in the country and far removed from medical help, and which they should grow for themselves: "*Carduus Benedictus, or the Bleſſed Thiſtle, is much vſed in the time of any infection or plague, as alſo to expell any euill ſymptome from the heart at all other times. It is vſed likewiſe to be boyled in poſſet drink, & giuen to them that haue an ague, to help to cure it by ſweating or otherwiſe. It is vſually ſowen of ſeed, and dyeth when it hath giuen ſeed.*"

Elsewhere Parkinson says of the plant: "*The diftilled water hereof is much vfed to be drunke againft agues of all fortes, eyther peftilentiall or humorall, of long continuance or of leffe: but the decoction of the herbe giuen in due time, hath the more forcible operation...*" [Paradisus Terrestris]

1665 – In *Panbotanologia* Lovell's explanation of the uses in plague are inter-twined with other attributes for the plant: "Bleffed thiftle, *Temperature is bitter hot and dry 2o* [ED – the second degree], *cleanfing and opening. Vertue. taken in meat and* drink *it help thofe that are vertiginous, deaf, or weak of memory. The fame boiled in wine and drunk hot, help gripings of the belly, killeth worms, causeth fweat, expelleth urine and gravell cleanfeth the ftomack, and help quartan feavers. The* juice *help all poison and inflammation of the liver: the powder of the leaves drunk in the quantity of* unc: fem. *helpeth the peftilence being prefently taken; fo the* decoction *of the wine. The green herb applied help all hot fwellings, as the eryfipelas, plague fores, and botches, the bitings of mad dogs or any venemous beafts; fo drunk as alfo the* diftilled water." [Panbotanologia, or Compleat Herbal]

1693 – Dale's Latin language *Pharmacologia*, aimed at Britain's medical profession, refers to Schroeder's German pharmacopoeia published in the 1640s for carline thistle virtues: "*Alexipharmaca eft & fudorifica, Peftis contagia arcet, curatque fi tempeftivè propinetur.*" [Pharmacologia]

1710 – Salmon's *Botanologia*, aimed at physicians, has the following for carline thistle: "The Liquid Juice. *It is given from one Spoonful to two or more, Morning and Night, againft the Plague, and Peftilential Difeafes, Meafles, Small Pox, Spotted Fever, Poifon... It may be given in Wine or fome other convenient Vehicle...*

The Decoction in Wine. *It has all the former Virtues, but as it is weaker, fo it muft be given from a quarter to half a Pint or more, Morning and Night, and fo to be continued for fome confiderable time.*"

Regarding *C. benedictus* Salmon refers to a hand-me-down remedy: "Roger Dixon's *Antipeftilential Sudorifick. Take Leaves of* Carduus Benedictus *and Scabious; Roots of Butterbur, of each 4 Ounces; Poffet Drink 3 Quarts; make a Decoction and ftrain out.* If you have not Poffet-Drink, you may make it with thin Water-Gruel. With this Drink, he told me, he Cured many Hundreds of the Plague (when the laft Great Plague was in *London, Anno 1665.*) who, by drinking largely of it, were Cured (many of them) in the fpace of twenty four Hours, when nothing but Death was expected..." *Benedictus* juice and the decoction were also recommended, and the green herb as a cataplasm for plague sores. [Botanologia]

1710 – For domestic readers Salmon's *Family Dictionary* has the following home remedy to expel the plague and its contagion, and although it does not specify *benedictus* that is the probable candidate for the *Carduus* ingredient: "*Take Rue, Wormwood, and Angelica tops, of each half a handful; Celandine, Carduus, of each* a handful and a half; put them into a Glazed Earthen Pot, when bruiſed together with a Pint of ſtrong White Wine Vinegar; ſtop the Pot cloſe, then let it ſeeth in Balnea Mariae, till the third part be conſumed, and then ſtrain it out, and keep it cloſe ſtopp'd. Let the Party infected, drink two or three Ounces of it, and Sweat after it, without Sleeping, a conſiderable time, if it may poſſibly be prevented; it Fortifies the Heart, aſſiſting Nature againſt Poiſons, and Infectious Airs." [Family Dictionary]

1714 – From a *Collection of Above Three Hundred Receipts*, aimed at everyday households (very much like the Beaton publications) there is the following recipe for 'A *good* Epidemick-Water' and perhaps *benedictus* is again present: "*Take Rue, Roſemary, Pimpernel, Roſaſolis, Balm, Scordium, Carduus, Dragon, Marigold-flowers and Leaves, Goats Rue, Mint, and of each two handfuls; take the Roots of Elecampane, Piony, Maſterwort, and Butter-bur, of each one pound; Gentian, Tormentil, Scorzonera, and* Virginia-*Snake-weed, of each four ounces; Saffron one ounce: Infuſe all theſe, when ſhred, in two Quarts of White-wine, and one quart of* French-*Brandy diſtil'd, and uſe it in any Malignant Diſtemper.*" [Collection of Above Three Hundred Receipts]

1714 – For *Carduus benedictus* the de Ville reprint of Caspar Bauhin's *Histoire des Plantes* follows along similar lines as English knowledge: "*Il eſt tout amer, enſuite il eſt chaud, apperitif, ſudorifique, & en grande eſtime par tout comme ſon nom le témoigne. La poudre des feüilles au poids d'une demi dragme beüe dans du vin eſt un excellent remede contre la peſte, mais il la faut prendre avant 24. Heures outre qu'elle conſume les flegmes de l'eſtomach.*" [Histoire des Plantes de L'Europe]

1727 – Strother's *Materia Medica*, referring to *C. benedictus* herbage, comments: "*If it be given in Decoction, it drives forth the febrile Matter thro' the Pores, or Urine, and is preſcrib'd before the Paroxyſm or Fit of the Fever. It is a Specifick in the Plague and Pleuriſy...*" [Materia Medica]

1747 – About the carline thistle James' *Pharmacopœia Universalis* says: "*The Root, which is the only Part uſed, is eſteemed ſudorific, alexipharmic, and efficacious againſt all contagious and peſtilential Diſeaſes, the Plague itſelf not even excepted.*"

   Regarding *Carduus benedictus*, which James names *Carduus luteus procumbens ſudorificus*, he quotes an account of the plant by German physician Hoffman (1660-1742): "*But its efficacy is chiefly celebrated in that formidable Diſtemper the Plague, againſt which it is uſed, both internally and externally. Internally it is exhibited, both with a preſervative and curative Intention; ſince it powerfully excites a Diaphoreſis. Externally it is applied, for breaking and opening Peſtilential Buboes, with which Intention it is alſo applied to other Impoſtumations. In the Opinion of the common People, a wine prepared of it in the Autumn is little leſſ than a Panacea, or univerſal Remedy.*" [Pharmacopœia Universalis]

1747 – The domestic-oriented *Medicina Britannica*, by Thomas Shore, sings many praises about the blessed thistle. Interestingly, he outlines its properties as both a vomit and sweat, which would have tallied with some plague treatment regimes: "... *its chief Praife is for the Plague, ufed inwardly to provoke Sweat either for Prevention or Cure; outwardly for breaking* Buboes *or other Impofthumes. If gathered in the Beginning of* June, *it is an excellent Vulnerary for green Wounds, but anfwers not that Intent any other time. The Decocotion of the dried herb in Poffet Drink, drank in fmall Dofes, provokes Sweat; in a large Dofe it vomits.*" [Medicina Britannica]

1751 – By the time of John Hill's professional *Materia Medica* in the middle of the 18th century carline thistle root appears to have been on the wane as a plague remedy among the local apothecaries' trade: "*The Root of the* Carlina *is efteemed fudorific, cordial, and alexipharmic: It is not kept in our Shops, but in fome other Nations it is as frequent in Prefcription on thefe Occafions, as Contrayerva is with us. Great Virtues are attributed to it alfo in malignant and peftilential Fevers.*"

In contrast to virtually all the previous attributed herbal qualities of *benedictus* the entry for that species would suggest that by the mid-18th century questions were surfacing over the validity of rehearsed herbal wisdom: "*Carduus Benedictus has been celebrated by the Writers of earlier Ages, as one of the greateft Medicines of the vegetable World. It has been recommended as an Alexiterial, Sudorific, and Cordial; and the Cure of Pleurifies, Peripneumonies, and intermittent Fevers, have been enumerated among the leaft of its Effects. It was much depended upon in malignant Fevers, and even in the Plague. Its Juice was fometimes given, fometimes a Decoction of it, and fometimes the dried Leaves in Powder. At prefent however thefe great Virtues are much fufpected, and the greateft Ufe made of it is by Way of Infufion for the working off an Emetic. Some indeed take this Infufion alone as a Vomit, and it will fometimes fucceed very well, but it requires large Dofes. The Seeds have been recommended in Emulfions with Almonds and other Ingredients, and fometimes alone in Pleurifies, and to promote the Eruption of the Puftules in the fmall Pox.*

We ufed to keep a fimple Water of Carduus Benedictus *in the Shops, but it has been found to have little or nothing of the Virtues of the Plant, and therefore has been of late wholly difufed.*" [History of the Materia Medica]

1754 – Regarding *benedictus* root Hill says in his domestic *Useful Family Herbal*: "*This is the only Part of the Plant ufed in Medicine. They fay it is a Remedy for the Plague: But however that may be, it is good in nervous Complaints, and in Stoppages of the Menfes.*" [Useful Family Herbal]

1814 – By the time of Thornton's domestic *Family Herbal* at the start of the 19th century it would appear that blessed thistle was out of fashion as the medical profession pursued chemical medicines: "*The virtues of this plant seem to be little known in the present practice. The nauseous decoction is sometimes used to provoke vomiting, and a strong infusion to promote the operation of other emetics... A stronger infusion, made in cold or warm water, if drunk freely, and the patient kept warm, occasions a plentiful sweat, and promotes the secretions in general.*" [Family Herbal]

BORAGE – *Borago officinalis* / BUGLOSS – *Anchusa arvensis*

Borage had a bit of a mixed reception over the centuries as an anti-pestilential herb and its inclusion within the list of plague plants here is really because it is such a well-known and familiar plant to herbalists and plant lovers. Generally borage is regarded as a garden plant although it is found in the wild; usually near to areas where there has been human habitation in the form of kitchen, cottage and herb gardens. With vivid blue, clustered flowers, this herbaceous biennial (sometimes annual) will grow about two to three feet tall.

Intensely bristly, to a point of feeling a little uncomfortable to grasp strongly, borage foliage and stems have the taste and aroma somewhat like cucumber flesh. In traditional herbal practice borage was ranked among the five emollient herbs, but by the mid-18th century was going out of favour. According to some past herbal wisdom borage leaves were employed by ordinary folks for healing flesh wounds and cleansing old ulcers, while writers such as John Gerard in the late 16th century claimed that borage could allay melancholia and raise the spirits. Other reported uses include a tea of the leaves against colds, use in kidney and bladder inflammations, and also rheumatism and respiratory troubles.

One problem that arises when associating borage as a former anti-plague herb is that the plant appears to have been used interchangeably with bugloss, while similar vernacular common names were applied to both species varieties; mainly *Lang de beefe* (Gerard) and *Langue de beuf* (Dodoens-Lyte). Gerard [1597] tells the readers of his herbal that *'Bugloffe is the true Borage'*, and: "*That which the Apothecaries call Bugloffe, bringeth foorth leaues longer then* [sic] *thofe of Borage, fharpe pointed, greater then* [sic] *the leaues of Beetes, rough and hairie,"* and relating to the entry directly for bugloss: *"Phifitions of the later time vfe the leaues, flowers, and rootes, in fteede of Borage, and put them both into all kindes of medicines indifferently, which are of force and vertue to drive awaie forowe and penfiueneffe of the minde, and to comfort and ftrengthen the hart."*

Waller [1822] tells his domestic readers that bugloss is: *"... so nearly allied to those of borage described above, that it will be by no means necessary to recapitulate them here. These two plants are generally found combined together in prescription, and one is frequently substituted for the other in practice".*

Today we recognize bugloss as *Anchusa arvensis*, related to alkanet and comfrey, but the Dodoens-Lyte and Gerard sources, in the view of this writer, do not appear to suggest that the related common alkanet (*Anchusa officinalis*) was used as yet another borage substitute despite Gerard calling alkanet 'Wilde Bugloss'. That said, Gerard refers to bugloss as used in agues [fevers, such as those associated with malaria] while the Dodoens translation attributes bugloss with similar anti-melancholy properties as borage. In regard to the latter attribute

Gerard says the: "*Sirrupe made of the flowers of Borage, comforteth the hart, purgeth melancholie, quieteth the phrenticke or lunaticke perfon.*" Given that the *symptoms* displayed by some plague victims could include delerium to the point of being comatose (possibly resulting from the secondary meningeal plague form), perhaps it was the phrenzy-reducing or calming aspects of the bugloss or borage species that encouraged some physicians to add them to their plague-remedies.

The reason why this all becomes significant in the case of tracking the borage-plague connection is that in the English context it appears quite hard to find many plague remedies where borage is a *key* ingredient, although water of borage appears as an admixture on occasion. If one looks through southern European 16th century plague texts, however, there is frequent mention on *bugloss* among plague remedies. Two works in the Bibliothèque nationale de France, for example, both repeatedly specify bugloss alone among curative ingredients; the Spanish text *Regimiento contra la peste* [Fernand Alvarez, circa late 15th century] and the Italian language *Contro alla peste* [Marsilio Ficino, 1523] respectively. Another late 15th century Latin plague text – *Tractatus de curatione pestiferorum apostematum* – by the Italian physician Rolando Capelluti, puts borage and bugloss side by side in almost every plague recipe where the plants are referenced. By comparison the northern European texts of Dodoens and Theodor Dorstenio's *Botanicon* [Frankfurt, 1540], from the same period as Dodoens, focus on the blood cleansing, melancholy and simple fever controlling aspects of bugloss and borage, with no hint of anti-pestilential properties.

1542 – In a chapter on what Andrew Boorde called the '*vsuall Herbes*', borage and bugloss head the list of edible and herbal plants: "*Borage doth comforte the herte, and doth ingender good blode, and causeth a man to be mery, & doth set a man in temporaunce. And so doth buglosse, for he is taken of more vygor, & strength, & effycacye.*" There is no mention of borage as an anti-plague plant nor is there mention by Boorde of borage in the anti-pestilential diet that he recommends, whereas Kellwaye [1593] and Thayre [1625] both list borage and bugloss in the dietary regime against pestilence. [Compendyous Regiment or a Dyetry of Health]

1569 – John Vandernote had the following recommendation for preservative dietary pottages 'when in feare of plague': "*The potage whiche fhalbe eaten in this time fhalbe made with wheate flouer, or ryfe, or lenttes. Howbeit, the firft water fhalbe caft away. Or els you fhall eate aleberries made with bere & bread or with red Cicers with ŷ fhels. The root of fenel of [sic] Smallach are good in pottage, & eaten, and all manner of meat made with egges, Save potaige made with dough or other comon potaiges are not good, faue potaige of Spinaige, or Percele, of Burrage, hartestonge, ar [sic] very good. All manner of herbes that haue a fondry vertue, as Rue, Scabiouffe, Ifop, Marubin, & fuche lyke.*" [Governance and Preservation of them that Feare the Plague]

1583 – Antoine Royet's French-language plague treatise includes borage, and bugloss, in both an plague electuary and in a potion to keep the Humours in balance. Under the heading of '*Electuaries fort proffitables*' there is the following recipe:

Rx. Theriacæ Alexādrinæ, 3.iij. Specierum lætitiæ Galeni Ɔ.j.ß. Boli Armeni, 3.ß. Terræ ſigillatæ Ɔ.vj. Conſeruæ roſarum, Bugloſi, Boraginis, an. Ɔ.iiij. Miſce Doſis. ℥.ß. in aqua ſcabioſæ, Angelicæ, & Boraginis."

To keep the humours in order (the reader may remember in an earlier section of this book there was mention of the medical belief system behind the concept of the *Humours*) Royet has a "*Potion pour diminuer doucement la quantité des humeurs ſans eſmouuoir*". The extended contextual text extract has been included here since the following paragraph also has a further reference to borage:

"*Rx. Summitatum lupilli, Fumarię an.m.ß. Capillorum Veneris, Boraginis an.m.j. Florū violarum p.j. Senæ Orientalis, Polypodij quercini recentis an. ℥.ß. Thamarindorum electorū 3 ij. Sem. aniſi 3 j.*

Macerentur per noctem in ſero caprino quātum ſuff. poſtea leuiſſimè bulliant: & ad colati iuris ℥ iij aut quatuor, adde ſyrupi roſati laxatiui ℥ j.

Pour la mondification du ſang eſt principalement recommandé le ſuc l'endiuie, cichorée, fumeterre, houbelon, bourrache, & millepertuis. Mais les Tamarins entre tous les fruicts peuuent deſtourner la putréfaction par leur aigreur." [Excellent Traicte de la Peste]

1587 – An interesting, and somewhat unusual, anti-plague remedy is the following one from Caspar Schvenckfelt's work, where borage and bugloss are teamed up in a 'water' of green nuts. Notably, balm (*Melissa*) also features, and the feeling that this writer has in trawling through old plague texts is that balm was more widely used in anti-plague remedies on the Continent than in the English context:

"*Aqva Nucum contra Pestem*
Rx. Rutæ, Scabioſæ, Acetoſæ, Borraginis, Bugloſſæ, Meliſſæ, Nucum viridium ana partes aeqvles: deſtillentur omnia ſimul in Balneo." [Thesaurus Pharmaceuticus]

1593 – In his *Defensative* Kellwaye recommends a particular 'preservative' from personal experience, '*An excellent good preſeruatiue; which I haue alwayes uſed with good ſucceſſe*':

"*Rx. Conſerue of Roſes and borrage flowers, of either two ounces. Minardus, Mitridate, Andromachus triacle, of either halfe an ounce. Pouder of the ſeede of Citrons pilled, one dramme. Syrop of lymons and ſower Citrons, of either halfe an ounce.*

Compounde all theſe together, in the forme of an opiat you may eate hereof euery morning, the quantitie of three beanes, and drinke a draſte of Renniſh wine, beere or ale, after it: but for Children and ſuch are of tender yæres, ſo much as a beane thereof is fufficient, and giue them only beere or ale after it: the taking hereof euery feconde or third day will fuffice, if you goe not into and fufpected company."

Regarding plague 'sweats' Kellwaye presents the following highly optional recipe that appears to allow the potion to be made up variably. The mention of 'small sorrel' there could be wood sorrel, or maybe it is sheep's sorrel (*Rumex acetosella*) and one wonders if 'bittaine' is betony:

"Take Conferve of the flowers of Borrage, Buglos, Violets, Bittaine, of either two ounces. Vennes Triacle, two ounces. Red Terra figillata, Terra lemnia, Mitridat, of either one ounce. Shauing of Eburni, And hartes horne, Orient Perles, Roots of Tormentill, of either one dramme. Shauing of vnicorns horne, Roote of Angelica, of either halfe a dramme. Syrop, of the luice of fmall Sorell and Buglos, of either fo much as fhall fuffice.

Mixe all thefe together, in the forme of an Opiat, then take of the fame Opiat, one dramme and halfe.

Scabios water, Balme water, of either two ounces.

Diffolue the Opiat in the waters, and drinke it warme, then walke a little upon it, and then goe to bed and fweate."

Following a purge, one of the natural courses of action for physicians dealing with plague, Kellwaye offers up *'A good Cordiall potion'* for post-purging:

"Rx. Aquarum bugloffæ, Acetofæ, ana. ℥. j.

Pul, diamarga, frigi, 3. B.

Take, Confectio alkermes, G. ij.

Syr, de aceto, Citri, Vel de limon, ℥. j. miffe.

All this you may take after purging... at any time... And here you muft underftand, that if it be in a plethoricke body full of ill humors, it were good that you purge him againe the next day." [Defensative against the Plague]

1665 – In a short publication issued by the Royal College of Physicians of London during the Plague of London the following sweat recipe appears under the heading of 'Expulsive Medicines': *"The poifon is expelled beft by Sweating, provoked by Poffet-ale, made with Fennel and Marygolds in winter, and with Sorrel, Buglofs, and Borage in Summer; with the which in both times they muft mingle* London-Treacle *the weight of two drams, and fo lay themfelves with all quietnefs to fweat."* [Certain Necessary Directions]

1693 – A hundred years later Samuel Dale refers to borage protecting against pestilential fevers and snake venom: *"Flores inter quatuor cordiales famofos recenfentur. Præparatis tribuunt vim roborandi cor, (& muniendi in febribus peftilentibus, & ictu ferpentum aliique venenis)... melancholicis excitandi, ac fanguinem depurandi.* [Pharmacologia]

1727 – In the 18th century one gets the feeling that the usefulness of borage is being questioned more often, and this is reflected in Edward Strother's translation of a Latin work by Paul Harman from Leyden regarding borage flowers: "*They are of a watriſh and inſipid Taſte, and do therefore conſiſt of watry Particles. Wherefore are they not cardiac, as has hitherto been ſuppos'd, becauſe not abounding with volatile oily Salts, which move forward the Blood, and add ſimilar Particles to our Spirits. But if they avail any way, it muſt be, that by their watry Particles they temper the Blood…*" [Materia Medica]

1751 – The tone of John Hill's *Lectures*, aimed at a more medically literate readership, also suggests that by the mid-18th century borage and bugloss were regarded as having dubious medicinal attributes: "*The Borage Flowers have always been eſteem'd Cordial, they are ſaid to inſpire Chearfulneſs, and prevent Faintings. There have not been wanting many who have carried this Encomium ſo far, as to talk of their curing peſtilential Fevers, and the Bites of venomous Animals, but this is idle. They are generally eſteem'd good in Pleuriſies, Peripneumonies, and all inflammatory Diſorders; and indeed ſo many things are ſaid of them by ſome Authors, which contradict the general Opinion of their Cordial Virtues, that it is not eaſy to ſuppoſe all true. On the whole, they are plainly enough of the number of thoſe Medicines which the preſent Practice very worthily lets ſink into Oblivion.*"

And of bugloss: "*Bugloſs is nearly ally'd to Borage in its Nature and Qualities. The dry'd Plant, thrown on a clear Fire, gives Marks of a nitrous Salt in the ſame manner as Borage does. The Flowers are eſteem'd greatly cordial: They are preſcribed by Authors, ſometimes in Form of Conſerve, ſometimes in Infuſion, and have much the ſame Effects attributed to them as are recorded of the other, but at preſent they are treated with the ſame Neglect.*" [History of the Materia Medica]

1770 – Relating to bugloss and borage flowers and foliage Alston's lecture notes comment: "*They may be eaten by pounds, like cabbage, any way. The conſerve is unknown here. They are but little uſed any where; and are excluded the* London M.M." [Lectures on the Materia Medica]

1790 – William Meyrick, writing for a domestic readership, talks in the past tense regarding some qualities of borage although he still seems to suggest that it is in use in other areas of herbal medicine: "*A water diſtilled from the flowers of this plant was formerly in great eſteem as a cordial, and ſtrengthener, but is very little regarded at preſent. It is however of a remarkable cooling nature, and conſequently may be uſed with ſucceſſ in inflammations of the eyes externally, and inwardly in burning fevers. The juice is good in obſtinate coughs, catarrhs, hoarſeneſs and defluxions on the lungs. The flowers made into a conſerve are ſaid to be good in putrid malignant fevers, and hypocondriacal complaints, likewiſe to remove obſtructions, and cure the jaundice.*" [New Family Herbal]

BURNET – *Poterium sanguisorba / Sanguisorba officinalis*

There are two 'burnets' commonly found in the wild: great burnet [*Sanguisorba officinalis*] and salad burnet [*Poterium sanguisorba*]. Both of these appear in herbal plague remedies, although salad burnet was perhaps more frequently used. In this respect it is useful to take note that both species were sometimes locally called Pimpernel, Pympurnalle, and Pimpinell, very much like *Anagallis*, so one needs to double-check any herbal reference to ascertain exact plant identity and physical details.

Salad burnet is mostly associated with chalky and limestone districts, particularly where the ground is sloping or hilly. Growing to a height of about two feet, the bruised foliage has the smell of cucumber and the bitterish taste of cucumber skin, and can be eaten in salads. In the latter half of the 18th century experiments were undertaken by the new generation of agricultural experimenters to transform perennial salad burnet into a perpetual livestock fodder, since the plant produces tender new growth a few weeks after it is cut back. For some reason, however, burnet never made it into the regular fodder crop favourites. On the herbal side, the plant has astringent qualities and was used in wound salves.

1570 – *Maison Rustique* is a 16th century French countryman's 'how-to' book on growing produce and raising livestock, but also deals with raising medicinal herbs for that added degree of self-reliance. Salad burnet is represented as a kitchen herb but the text also includes some brief notes on the medicinal use of the plant: "*La pimprenelle des jardins, qui eſt celle que lon met aux ſalades... Lon en fait grand cas en temps de peſte, & dit-on que le frequent vſage de pimprenelle eſt ſouuerain preſeruatif contre les maladies dangereuſes.*" [Maison Rustique]

1582 – In the late 16th century French plague text by Esaie le Lievre (perhaps a pseudonym) the writer suggests that where an individual patient becomes delerious – '*auct poinct de frenſye*' – then a topical epytheme, liniment and embrocation may be used. In the recipe which is given it is really quite hard to determine whether pinpinelle (burnet) or the pimpernell of *anagallis* is being specified, but a couple of paragraphs ahead of this text extract the author lists a number of dietary herbs among them pinpinelle, so perhaps we are looking at the cucumber-tasting burnet in this instance: "*Rx. Eaue de Bugloſe, D'ozeille, de Roze, de Chardō benict, de Calendulle, de Pinpinelle, de chacun, onc. ii, Camphre, drac. ſ. Safren oriental, pouldre froide de Margueritte, Diarodon, de chacun, drac. ii. Muſqz fin, Ambre gris, Ciuette, de chacun. ſcrup. ſ. de tout emſemble ſoit faict Epiteme.*" [Epydiomachie ou Combat de la Peste]

1629 – Parkinson's entry for burnet in his *Paridisus Terrestris* heads up the description of the plant as '*Pimpinelle ſiue Sanguisorba*' and, after mentioning that the plant can be used as a foodstuff, comments: *"It is alſo vſed in vulnerary drinkes, and to ſtay the fluxes and bleedings, for which purpoſes it is much recommended. It hath beene alſo much commended in contagious and peſtilentiall agues."* [Paradisus Terrestris]

1640 – A decade later in his *Theatrum Botanicum* Parkinson who, it should be noted, was an apothecary-botanist, seems to use juiced burnet as a 'sweat' ingredient: *"... it is a ſpeciall helpe to defend the heart from noyſome vapours, and from the infection of the Plague or Peſtilence, and all other, contagious diſeaſes, for which purpoſe it is of great effect, the juice thereof being taken in ſome drinke, and they either layd to ſweate thereupon, or wrapped and kept very warme."* [Theatrum Botanicum]

1641 – A contemporaneous plague text from Thomas Sherwood advocates pimpernell in a remedy for children during plague; *scabies* being scabious, and covered elsewhere here, while angelica has already been elaborated upon: *"Alſo the waters of* Scabies, Angelica, *or* Pimpernell, *are great preſervations for children againſt the Plague, if two ſpoonfuls thereof be taken in a morning faſting."* [Charitable Pestmaster]

1707 – In a French language reprint of Caspar Bauhin's botanical work *Histoire des Plantes* – by the publisher Nicholas de Ville, in Lyon – the text identifies *'pimpinella saxifraga major'* as an anti-plague herb: *"... le jus de la racine beu en vin, eſt ſingulier contre tous poiſons, & morſure de bête venimeuſe, & par ce moyen quelques uns en ſont grande eſtime contre la peſte."* Elsewhere the text quotes Matthiolus regarding *Poterium sanguisorba* as: *"... fait grande cas de cette herbe és Fievres peſtilentielles & contagieuses."* [Histoire des Plantes de L'Europe]

1710 – Reading some of William Salmon's works, which also included an almanac, one has the impression that he was something of a self-publicist despite being a Doctor, though some of his contemporary detractors were less polite on that account. In his *Botanologia* there are three references to burnet in the plague:

"The Vinegar. It is a ſpecial thing to preſerve from, and cure the Plague or Peſtilence, the Spotted Fever, or any other malign acute Diſeaſe of that kind; for it in a ſpecial manner defends the Heart from Poiſonous and Noiſom Vapors, and all contagious Diſeaſes, being given mixed with the Juice in equal quantities, and the Patient laid to Sweat thereupon. Doſe three ounces.

The Acid Tincture. *This has all the Virtues of the Vinegar aforegoing, but is much more powerful than it, to all the purpoſes and intentions there ſpecified: beſides this* [it] *is an extraordinary Stomatick, and therefore powerful in ſtopping Vomitings, and alſo ſpitting of Blood, being given in a Glaſs of Canary, or Red Port Wine. Doſe thirty or forty drops, to fifty or ſixty, two or three times a day.* [ED – 'Canary' was a popular sweet wine from the Canary Islands at the time.]

"The Spirit. ... *It prevails alſo againſt the Contagion of the Plague. Doſe from one dram to half an ounce.*" [Botanologia]

1710 – For domestic use Salmon's *Family Dictionary* provides its readers with the advice that burnet "*... preſerves againſt the Plague, and the Bitings of mad Dogs, and alſo reſiſts Poiſons*" then provides instructions for a 'water' that families could have made for themselves: "*Burnet Water: Take the Tops of Burnet twelve handfuls: Tops of Wormwood, Roſemary and Mugwort, Sun-dew, Dragons, Scabious, Agrimony, Carduus, Betony, Bawm* [sic]*, and the leſſer Centaury, of each a handful; Roots of Angelica, Peony, Zedoary, Tormentil, Liquorice and Elecampane, of each half an ounce; bruiſe ſhred and infuſe them with Sage, Rue, Celandine, Marigold leaves and flowers, of each a handful, three or four days, in four quarts of the fineſt White Wine, then diſtil it carefully. To the diſtilled liquor put freſh Burnet twelve handfuls; Sugar four pounds; digeſt ſix, eight, or ten days, then draw off the Liquor, ſo will you have an excellent Burnet Water.* Let the Doſe be 3 or 4 ſpoonfuls at a time." [The Family Dictionary]

1713 – In republishing Bate's Pharmacopœia – *Pharmacopœia Bateana* – for the herbal and medical trade Salmon adds his own comments and suggestions to Bate's work. In the following recipe for a cordial bole several of the plague plants listed elsewhere here contribute towards the ingredient list in addition to burnet:

"*Bolus Alexiteria, A Cordial Bole, or Troches reſiſting Poyſon.*

Bate.] *Rx Anglica* [sic]*, Bawm* [sic]*, Scordium, Burnet, Scabious, A.M. ij. Dragons, M. j. Flowers of Marigolds, Clovegillyflowers, A.M. iij. Wood of Aloes ʒſs. Berries of Kermes ʒiſs. Cochinele* 3*ij. White Wine* ℔*ij. vel q. s. digeſt warm for 24 Hours, then expreſs out, and with fine Bole in fine Powder* ʒ*viij. make a kind of Pulp, or Pudding, like Subſtance, which dry in the Sun. This Work repeat five times with the ſaid Infuſion, and at laſt form them into troches, S.A.* Doſe, 3*j. ad* 3*j.* in malign Diſeaſes.

Salmon.] I ſhould rather chooſe to make it with the Juices of the eight firſt Ingredients than with their Infuſion in White Wine, for thereby the Medicine would be both ſtronger and better.
It may be given Morning and Evening, as a Preventative againſt the Infection of the Plague, or the Contagion of other malign Diſeaſes." [Pharmacopœia Bateana]

1747 – Thomas Short's domestic-oriented *Medicina Britannica* highlights burnet for cordial and alexipharmic properties: "*Burnet* (Pimpinella) *is either Cordiall or Alexipharmac; hence infuſed green in Wine, it gives it a fine Aromatic Flavour and Taſte; it chears* [sic] *the Heart, and reſiſts the Plague, and contagious Diſeases...*" [Medicina Britannica]

BUTTERBUR – *Petasites hybridus*

In its young growth phase the leaves of perennial butterbur look a little like those of coltsfoot and it grows in similar sorts of habitats, although butterbur tends to have more of an affinity with moist soils so is often found growing by the sides of streams, canals and river banks. Moist clay soils, in particular, are favoured, but also moist sandy loams.

Once butterbur has matured, however, then the differences with coltsfoot become very obvious. For a start the flowers are not bright yellow on single stalks but appear, instead, as clustered tufts of small lilac-pink flowers which bloom around March to May. The species is *dioecious*, with male and female flowers borne on different plants which, in some parts of the British Isles, has resulted in areas where the female plant is virtually absent. The large flat leaves, also, develop a much more distinctive rhubarb-like heart shape profile, and are rather downy in appearance and to the touch. Lastly, in overall size the butterbur will grow a couple of feet tall in optimum conditions, while coltsfoot rarely reaches more than a hand length in height.

While some of the common local names for butterbur reflect its liking for moist habitats – Bog Rhubarb and Water Docken for example – other names are more obscure: Cap Dockin, Burn-blades, Kettle Dock, Dunnies, Flapper Dock, Flea-dock and Poison Rhubarb among them. Country children would sometimes call the plant Bogs Horns, which relates to youngsters using the hollow stalks as horns or trumpets. The name Butter-bur, and possibly Batter Dock too, derived from country folk wrapping butter in the large leaves. Most importantly, at least from the herbal perspective, it was called Pestilence Wort as butterbur was believed to be a remedy for plague and pestilential fevers. There are other species of petasites, and it appears that sometimes these were used as alternative substitutes, though the key one is *P. hybridus*.

1578 – The Lyte version of Dodoens' Flemish herbal says: "*Butter Burre dried, and made into powder and than* [sic] *dronken in wine, is a ſoueraigne medicine againſt the Plague, and Peſtilent feuers, bycause it prouoketh ſweate, and for that cauſe it driueth from the harte all venim, and euill heate.*" [New Herbal]

1640 – In *Theatrum Botanicum* apothecary-botanist John Parkinson almost mirrors the previous sentiment some eighty years later: "*The rootes hereof are hot in the firſt degree, but as* Galen *ſaith drie in the third, and are by long experience found to be very auailable againſt the plague, and peſtilentiall fevers by prouoking ſweat, if the powder thereof be taken in wine; as alſo reſiſteth the force of any other poyſon...*" [Theatrum Botanicum]

1665 – Regarding 'Butter-burre Petafitis' Lovell's Panbotanologia has another variation, that of being mixed with ale: "Vertue. the roots ſtamped with ale and drunk helpeth peſtilent and burning fevers, coole and abate their heate, poudered and drunke in wine helpeth the plague, causeth ſweat, and driveth from the heart all venome and ill heate...

... Aqua petaſitidis compoſitae Pharm. Lond. or, the Compound water of butter burre, is good with fit cordials againſt peſtilentiall feavers; cochl. I. Taken in the morning may be a prophylacktick." [Panbotanologia, or Compleat Herbal]

1707 – The early 18th century French version of Bauhin's Histoire des Plantes notes that the Germans name butterbur 'plague root': "Sa racine priſe en poudre dans du vin, au poids de deux dragmes, faiſant en ſuite fuer le malade, eſt un remede experimenté contre la peſte & les fievres peſtilentielles; auſſi les Allemans appellent cette racine, la racine de la peſte, la même potion ſert aux ſuffocations de la matrice." [Histoire des Plantes de L'Europe]

1710 – William Salmon's Botanlogia, aimed at practising medics, provides his readers with five options of using butterbur in plague: "It has a peculiar power and force againſt Poiſon of all kinds, and the Infection of the Plague.

The Pouder. Being given one dram, and drunk in Wine, it is a moſt excellent thing againſt the Plague, and all ſorts of Peſtilential Fevers... and drives from the Heart all ſorts of Venom and Poyſon...

The Juice. It is expreſſed out of the Root by being beaten in a Mortar, and ſqueezed out with White Port Wine. It has all the Virtues of the Pouder, and may be given from one ounce, to two or three ounces, Morning and Evening, as a preventive againſt the Plague, and every ſix hours after Infection.

The Eſſence. More powerful than powder and Juice and 'Doſe one to two ounces mixed with Wine.'

The Decoction. Doſe from two ounces to four or ſix, as hot as it can well be taken, Morning and Evening, to prevent the Infection of the Plague; and as much every ſix hours, for thoſe who are already ſeized with it.

The Mixture. It is thus made: Take of the Juice expreſſed with Vinegar, or the Eſſence, twelve ounces: Juice of Rue four ounces: Venice Treacle, or Mithridate two ounces: mix them well together by ſhaking. It is an Antidote againſt the Plague or Peſtilence... Doſe two ounces at a time, as often as need requires." [Botanologia]

1722 – Manget has this plague recipe for poor patients: "Pour les pauvres gens, on peut faire un petit laict avec vinaigre peſtilentiel, & ayant cuit dans ce petit laict la racine de petaſités, en donner à boire chaudement un trait." [Nouvelle Reflexions]

1747 – From Robert James' extensive Pharmacopœia Universalis the following extract reflects the use of two types of butterbur: "Petafites major & vulgaris,

Butter Bur. It grows in watery Places, flowering in March. *The Root is uſed, which is eſteem'd ſudorific, alexipharmic, and good in the Plague. It is recommended in hyſteric Fits, Coughs and Aſthma's. It kills the flat Worms in the Inteſtines, and excites Urine and the Menſes. Externally apply'd, it is good for* Buboes and malignant Ulcers."

James also relays details of an '*Aqua Epidemia* Plague Water' that appeared in the 1744 *Edinburgh Pharmacopoeia* a few years before, although from his follow-on comments he seems to have doubts about its efficacy: "*Take of the Roots of Maſterwort and Butter Burr, each four Ounces;* Virginia *Snake Root and Zedoary each two Ounces; the Seeds of Angelica and Bay-berries, each three Ounces; the Leaves of* Scordium, *ſix Ounces; bruiſe and cut the Ingredients, and pour thereon two Gallons of* French *Brandy; and when they have ſtood to macerate for four Days, draw two Gallons.*

This is ſaid to be intended as a highly carminative Cordial, in very low and languid Caſes, and to raſe [sic] *the Spirits in the Plague and malignant Fevers. But in theſe Caſes I ſhould ſuſpect it of doing great Miſchiefs, for the very ſame Reaſon that it gives a temporary Relief.*" [Pharmacopœia Universalis]

1747 – Short's domestic *Medicina Britannica* describes butterbur for domestic readers as: "*... bitter, but not hot, the Roots bark'd and ſteep'd in Vinegar, till it is impregnated with their Virtue, then drank with Juice of Rue and Treacle, are good in peſtilential Fevers... The Roots are Sudorific and Alexipharmac, good in Fevers, malignant, peſtilential, and contagious Diſeaſes...* [and] *are applied in Poultiſes to Buboes, and Plague Sores.*" And in a selection of plague remedies in the section on scordium Short quotes Hemmings for the following snippet: "*... take often a Dram of powdered Butterbur Root; or infuſe Rue in white Wine... or boil Rue in white Wine, and drink the Decoction with a little Vinegar.*" [Medicina Britannica]

1751 – The picture painted of butterbur in John Hill's *Materia Medica* is much wider, and he also picks up on the rustic use of butterbur for dealing with worms: "*Butterburr Root has at all times been famous in malignant and peſtilential Fevers. The* Germans *are ſo fond of it on theſe Occaſions, that they have named it Peſtilence Wort; and it has with us been made the Baſis of many cordial Waters and other Preparations intended againſt Contagion. It is certainly a very powerful Sudorific. It promotes alſo Urine and the Menſes, and it by many greatly recommended in Aſthmas, Coughs, and Difficulty of breathing. It is given alſo as an Aſtringent in the* Fluor albus, *and in old Gleets. The general Way of preſcribing it is in Decoction. The Country People uſe it againſt Worms, and apply it externally to Ulcers, which they ſay it cleanſes and heals.*" [History of the Materia Medica]

1790 – William Meyrick's domestic herbal provides its readers with a considerably longer account that reassures them that butterbur root could be used herbally '*for the worſt kind of fevers*' which, obviously, could conjure up plague fever in the worst case scenario. At the very bottom of the text Meyrick accredits John

Hill for some or all of the information that follows, but this writer has not been able to find the Hill source. Here's Meyrick's homespun advice:

"The method of uſing the root is this; After having cut away the fibres from the body of the root, and waſhed it, ſlice two ounces of it thin, into a clean earthen veſſel, and pour on it a quart of boiling hot ſoft water; let it ſtand till cold, and then pour it clear off, add about a quarter of a pint of mountain wine to it, and a little fine ſugar, and let a quarter of a pint of this be taken every fourth hour: the ſpirits will be raiſed, the anguiſh and depreſſion which accompanies theſe kind of fevers, and proclaim their fatility [sic]*, will be removed, a kind gentle ſweat will ſoon come on, every bad ſymptom will vaniſh, and the patient will ſpeedily recover his health.*

If in the worſt of caſes a boil or bubo ſhould make an appearance under the arms, or in any other part of the body, make an ordinary poultice of white bread and milk, and to a half pint baſon of it add a quarter of a pound of Butter-bur, roots bruiſed, but not boiled, together with a little ſallad oil. Apply this on the part; let it be kept warm, and renewed frequently, ſo will the patient have all the chance for a recovery which the nature of the caſe admits. I could ſay much more on this ſubject, but it would be an unneceſſary talk to prove that the ſun gives light, and it is no leſs certain that this root is the beſt known remedy for putrid and peſtilential fevers. Hill." [New Family Herbal]

1822 – Martin de Saint-Genis' early 19th century plague text features the following recipe for plague pills which had appeared in a re-work of Richard Mead's *Some Observations Concerning the Plague*, published one hundred years before in 1721:

"Pilules anti-pestilentielles de Diemerbroeck, pour entretenir la liberté du ventre. PRENEZ racine de petasite, de carline, de dictame, d'angelique[,] *d'aunée, de chaque, demi-once, de gentiane une drachme et demie, de belle rhubarbe une once et demie, de l'agaric bien blanc une demi-once; des herbes de scordium, de petite centaurée, de rhue, de chaque, une demi-once; de chardon-béni six drachmes; de fleurs de staechas une drachme et demie, des semences de citron, d'orange et de la zedouaire de chaque une drachme.*

Faites de tout cela une poudre grossière que vous ferez macérer pendant deux ou trois jours dans deux livres et demie où trois livres de vin blanc; cuisez ensuite pendant un quart d'heure, et coulez et exprimez; passez cette colature par le papier gris, et y dissolvez ensuite aloès, succotins trois onces et demie, myrrhe en larmes pures trois drachmes et demie. Faites évaporer à feu doux jusqu'à consistance d'extrait pour en former des pilules." [Manuel Préservatif et Curatif de la Peste]

ORANGE

LEMON

COMMON ELDER

ELECAMPANE

CLOVE-GILLYFLOWER

JUNIPER

MARIGOLD

MASTERWORT

CITRUS, LEMON – *Citrus limon* / ORANGE – *Citrus aurantium*

The use and incorporation of citrus fruits into west European medicine very much follows the history of their introduction into various regions. For many centuries citrus was regarded as highly exotic and had a high price tag to match; however oranges, lemons and limes were quickly recognized for their use in herbal remedies, and in the treatment of plague too. In earlier centuries the citron (*Citrus medica*) – a gnarly relative of the citrus fruits we know today – was mostly employed, until lemons and oranges became more widely distributed. The sharp acid *hit* of lemons appears to have been preferred, while orange was viewed as having softer and more delicate medicinal qualities. Of orange species, it was the bitter Seville fruit (*C. aurantium*) that was preferred, while Orange trees appear to have been grown in Britain from around 1629.

Although the text of Edward Strother's *Materia Medica* [1729] does not specify orange peel for treatment of the plague, his description of the virtues of oranges provides an insight into how physicians of the period perceived the beneficial qualities of orange peel, the juice also, and the pulp as a preserver of the memory: "*It conſiſts of volatile and oily Particles, join'd along with ſome terreſtrial Ones alſo. From whence it incides, opens and attenuates, by reaſon of its volatile, and oily Salts, which, being join'd to ſome earthy Ones, makes it a Specifick in Diſeaſes of the Stomach and Inteſtines; becauſe by its volatile Parts it incides the viſcid Matter, and ſheathes it by its oily Ones; as it corroborates the Fibres, and ſo fortifies them, that they may expel it more forcibly... We preſcribe its exterior Part, preſs'd out by our Fingers, as a Preſerver of the Memory, an exhilerating Cordial, as a Cephalick, and an Alexipharmack, as a Remedy for a Syncope, or a Swooning; as an Antiſcorbutick, where the Matter is viſcid and acid; as we do its Juice for a Cooler.*"

1570 – Estienne's *Maison rustique*, which is essentially a French countryman's and farmers book, suggests that its readers consider growing citrus among other fruit trees, but mentions the potential medicinal benefits of citrus too including their use in plague: "*L'eſcorce, le ius, la ſemence de citron font tous ſouuerains contre toute ſorte de poiſon & danger de peſte: autant en eſt-il du limon. Et pour ceſt effect lon peut faire cuire vn citron & limon tout entier en eau roſe & ſucre, iuſques à tant qu'il ſoit du tout conſumé en ius: puis vſer tous les matins à la quantité d'vne ou deux cueillerees de ceſte decoction en temps de peſte.*" [Maison Rustique]

1575 – From the 1649 English version of Paré's work we find him recommending that patients suffering from a burning plague fever should avoid drinking wine, and drink oxymel instead. Failing that he suggests the following sugared water: "*Take two quarts of fair water, of hard ſugar, ix ounces; of cinnamon, two ounces; ſtrain it through a woollen bag or cloth without anie boiling; and when the patient will uſe it, put thereto a little of the juice of Citrons. The ſyrup of the juice of Citrons excelleth amongſt all others that are uſed againſt the peſtilence.*"

A little later Paré describes the following thirst-quenching juleps, with citrus again extensively employed: "*Take of fair water, one quart; or white or red vineger* [sic] *three ounces; of fine fugar, four ounces; of fyrup of Rofes, two ounces: boil them a little, and then give the patient thereof to drink. Or take the juice of Limons and Citrons, of each half an ounce; of the juice of fowr Pomgranats, two ounces; of the water of Sorrel and Rofes, of each an ounce; of fair water boiled, as much as fhall fuffice: make thereof a Julip, and ufe it between meals. Or take the fyrup of Limmons and of red currance, of each one pound; of the water of lillies, four ounces; of fair water boiled, half a pinte* [sic]*; make thereof a Julip.*" [Workes of Ambrose Parey]

1578 – Regarding the citrus fruits oranges and lemons, the Dodoens-Lyte herbal from the Brabant region of Belgium says: "*The iuyce of thefe fruites, and the inner fubftance wherein the iuyce if contained, efpecially of the Orenges, is very good againft contagioufneffe and corruption of the ayre, againft the plague & other hoate feuers, and it doth not onely preferue and defende the people from fuche dangerous fickneffe, but alfo it cureth the fame.*" [New Herbal]

1583 – With close access to the citrus fruit growing areas of the Mediterranean it is perhaps not surprising that many medical texts from that region of Europe feature citrus in anti-plague remedies. The following extracts are from Royet's French plague treatise, the first under the heading '*Compofitions preferuatiues*':

"*Rx Corticum citri, & Mali aurei faccharo cōdito an. ℥ j. Conferuæ rofarum, Rad. bugloffi an ℥ iij. Sem.citri ℈ iij. ß. Sem.anifi, Fœniculi an. ℈. ß.*
Rad. Angelicæ ℈ iiij. Sacchari rofati quantum fuff. Fiat conditum coopertum foliis aureis, quovtatur ex cocleari, vt dixi, in exitu domus."

In a section entitled '*Remedes, deffenfifz, & preferuatifz*' Royet also seems to suggest citrus forms part of the dietary regime during hot weather: "*Quand le temps fera fort chaud, il faudra prendre tous les matins vne mye* [sic] *de pain trempee dedans du ius de limons, ou d'orenges, ou bien en vinaigre, & eau rofe, ou vn bouillon de poullet cuit auec ozeille. Et pour eftre meilleur y adioufter canelle, & fuccre.*"

1585 – Italian physicians had similar easy access to citrus fruits and we find the following in Durante's herbal printed in Rome: "*Fafsi del fucco de i limoni, cofi de i cedri, vn firoppo vtile à fpegnere la caldez za della colera, & nelle febri contagiofe, & peftilential.*" [Herbari Nuovo]

1590 – Barrough's *Method of Phyfick* suggests a potion for strengthening the heart at the time of need in a plague attack: "*... you muft minifter medicines (fpecially if the ftrength be feeble) which can ftrengthen and comfort the heart, and other principall members of the bodie, as is this: Rx. Conferues of Violets, Rofes, and Bugloffe, ana ℥.j.ß Bolearmoniacke prepared, 3.j. red Corall, ℈.j barkes of Citron apples, 3.j.ß Camphire, ℈.v with firupe of the iuyce of fharpe Citrons, as much as is fufficient, make an Electuary or liquid Antidote.*" [Method of Physick]

1614 – In a French domestic medical work from the early 17th century we have the following anti-plague *Potion pour les pauures* which rather suggests, given the proximity of France to the orange growing regions of the Mediterranean, that medicines made with citrus would not be beyond the reach of ordinary folks: "*Suc d'ozeill bien clair, trois onces d'orenges aigres, bon vinaigre blanc & clair, eau rofe de chacune vne once. Faites bruuage, en adiouſtant vn peu de ſuccre qui voudra.*" [Les Secrets du Seigneur Piemontais]

1614 – In Vigier's short plague book *Traicte de Peste* citrus is among the recommended good foods to eat to keep healthy in plague times: "*Les ſalades de citrons, limones, & oranges auec eau rofe & ſucre font bonnes, comme auſſi l'vſage des grenades, oliues, cappres & fenouils marins.*" [Traicte de Peste]

1625 – Thomas Thayre's plague treatise lists an anti-plague preservative pill recipe but warns that these pills, and some others, were not recommended to be taken during pregnancy and so recommends an alternative diet with a citrus component as follows: "*But women great with child may not take of theſe pils, neither of the other pils ſet downe before: let them content themſelues to eate in a morning, ſome conſerues of ſorrell, roſes, or borrage, wherewith they may mixe ſome ſirrup of Limons, and let them be mery and vſe a good diet, and good company to paſſe the time away, and this is the beſt medicine I can aduiſe them.*"

On a very general level Thayre's advice on diet in pestilential times suggests: "*The vſe of Orenges, limons, and pomgranats, is very good; ſo is Vineger, Cloues, maces, ſaffron, ſorrell with your meat, or* either of them in a morning with ſugar is good. Let all your meates bee dreſt and ſauſed with Vineger, Orenges, and Limons, Maces and Saffron, and a little Cinnamon, and auoide all ſtrong wines, and hot ſpices.* [* That is, Orenges, Limōs, Poungranats.]*" [Excellent Treatise on the Plague]

1636 – In Stephen Bradwell's *Physick for the Sicknesse, Commonly Called Plague*, which was written about a decade after the 1625 plague outbreak in London, and was one of many plague and pestilence advice pamphlets written by physicians and apothecaries promoting their own services in perilous times, we have the following recommendation: "*Carrie in your mouth a peece of* Citron-pill, *or for want of that, of* Lemon-pill; a Clove, *or a peece of Tormentill Root. Or if any will reſort to me in* Golding Lane, *I will ſoone provide for them* Lozenges *to hold in their mouth, fit for their conſtitution, and ſuch as I have had good experience of, the laſt great Plague time.*" [Physick for the Sickness]

1703 – The early 18th century translation of Etmuller's *Practice of Physic* brought German medical thinking to British shores; several editions of the family orientated *Etmullerus Abridg'd* appearing around the turn of the 17th-18th centuries. The section on *Plague and Peſtilential Fevers* is just the sort of information that a family might have turned to when news of the 1720 plague outbreak in Marseilles reached them. There are Two recipes that contain orange, or other citrus ingredients, as follows:

"*The Specifics that are us'd in this Difeafe, are Camphyr given in fubftance, or its Oyl mixed with Oyl of Amber, and that of Citron Peel, called* Heinfius's *Oyl; Ivy-berries, given to a dram in Vinegar and Wine; Juniper and Elder berries, and the Rob of either given in Vinegar; Garlic heads brais'd and exhibited in Vinegar; the Blood of a Stork, or its volatil Salt...*

Take of the Water of Carduus Benedictus, an ounce and a half; Vinegar of Wine, fix drams; Diafcordium, a dram and a half; Camphyr, fix grains; Syrup of the Juice of Citrons, half an ounce. Make a Potion for one or two Dofes." [Etmullerus Abridg'd]

1710 – William Salmon's domestic *Family Dictionary* mixes citrus components with some of those other common herbal ingredients such as rue and angelica: "Bezoartick Balfam: *Take diftilled Oils of Rue, of Citrons, of Oranges, of Lavender, and Angelica, of each a fcruple; Oil of Amber rectified five Drops, Camphire ten Grains, Oil of Nutmegs half an ounce; make thefe into a Balfam by well incorporating over a gentle Fire...* It is good in Peftilential Airs, and Apoplectick Fits, or any Diforder of the Brain."

Elsewhere Salmon informs his domestic readers of the general benefits of citrus fruit in plague and offers up a recipe to make orange syrup at home:

"*Citron: The Juice of it repreffes Choler, (and if made into a Syrup with fine Sugar) is very good againft the Plague and Peftilential Fevers...*

Citrons: a Syrup, Take Citrons as many as you think convenient, pare and flice them very thin, then put them into a Silver Bafon with Layings of fine Sugar, till it be near full; the day following pour off the Liquor into a Glafs, with a Glafs Funnel, ftrain it through a fine Strainer, clarify it on a foft Fire, and it will keep a [sic] *twelve Month. This is excellent in hot Diftempers, efpecially mixed with Juleps and Cordials.*

Orange Water... is very good in Peftilential Fevers: It ftrengthens the Heart and the Brain. Three or four fpoonfuls taken going to Bed, caufes likewife a gentle breathing Sweat: The Juice of Oranges is cold, and therefore refifteth Corruption, and is given with a little Sugar, fuccefsfully to cool and temperate the Blood in Fevers and hot Difeafes."

Salmon then gives three methods of making the orange water, this one among them: "*Take two quarts of Sack, a pint of Brandy, the Rind of twelve Oranges thin pared, two Drams of Saffron; fteep thefe together two Days, then Still them in a Cold Still, put fome fine Loaf Sugar and Leaf Gold, into the Glafs the Still drops into, with three Grains of Ambergrife, fweeten it to your Tafte; you may draw two quarts of Water from this quantity.*" [Family Dictionary]

1722 – Despite Salmon's enthusiasm for orange in plague cures, there appears to have been doubt in the minds of some practitioners as Quincy's *Pharmacopœia* suggests regarding *Aurantium*: "*They are now but little uſed in Medicine: However, where the Spirits are almoſt quite extinct in malignant and putrid Fevers, they are ſometimes very cordial and refreſhing.*" [Pharmacopœia Officinalis]

1758 – A glimpse of mid 18th century domestic home remedies comes from the pages of Lydia Honeywell's rare *Cook's Pocket Companion & Universal Physician*. The recipe is for a 'water' for living to an old age but the author adds that: "*It will prevent Infection ſeizing the Heart in a Time of Peſtilence...*" among other conditions that it could be used for. In addition to citrus peel the recipe has a significant rose content, some expensive spices, and large quantities involved: "*... take three Quarts of Roſe-water, ten Ounces of Orange and Lemon-peel dry'd in the Shade; Cinnamon, Cloves and Nutmegs, of each half a Pound; Red-roſes that have not been gathered more than two Days, two Pounds, four pinches of Roſemary-tops, and two of Laurel-leaves, four Handfuls of Marjoram, as many of Balme gentle, four Pound of Hyſſop, as many of wild Roſes.*
    Put all theſe together with Roſe-water, bed up on bed, into a Glaſs-alembic, and then diſtil them very gently in a Balneo Mariæ *or* Bath Mary, *and keep the Water that comes out for uſe... the Doſe is about two Spoonfuls, Morning and Evening; and rub any diſordered Parts with it.*" [Cook's Pocket Companion]

1790 – By the time of William Meyrick's domestically oriented *Family Herbal* citrus fruits would have been more accessible to many sections of the population, though perhaps not the poorest in society. The passage below, regarding oranges, shows how citrus was generally seen as healthy and useful in remedies: "*The yellow rind of this fruit, when carefully freed from the white fungous matter underneath, is a grateful, warm, aromatic bitter, of great uſe as a ſtomachic and corroborant, and for giving an agreeable taſte to other medicines, being warmer than the peel of lemons, and of a more durable flavour. The juice of oranges is a grateful acid, and of great uſe in both putrid and inflammatory diſorders.*" [New Family Herbal]

COMMON ELDER – *Sambucus nigra*

There can be few people interested in wild plants, or who have a love for the countryside, that are unfamiliar with the common elder, and even those who are personally unacquainted with the shrub will undoubtedly have heard of elder-flower champagne or elderberry wine.

Elder is a common tree of hedgerows and roads, and is frequently found near to human habitation in country districts as elder was commonly used in herbal remedies and was supposed to fend off evil spirits too; although in the British tradition elder was regarded as a tree of ill omen so you did not want a specimen growing too close to your home despite the approved herbal uses, about which Thomas Green [1823] quaintly put it: "*This tree is, as it were, a whole magazine of physic to rustic practitioners, nor is it quite neglected by more regular ones.*"

Elder is generally found as a shrub but, given enough time to grow, can reach heights of twenty to thirty feet, taking on the habit of a tree. The flat-topped conspicuous masses of small, creamy-white, strong-scented flowers, that appear around May to June eventually give rise to clusters of small, purple-black, berries. Herbally, the inner green bark was once used for dropsy, while the flowers are diaphoretic and expectorant. The berries were used for their cordial, diuretic, and anti-dropsical actions, and also as a gargle for sore mouths and throats.

The common names for elder are highly varied, with Alderne, Arntree, Baw-tree, Boortree, Bore, Devil's Wood, Eldern, Hydul-tree, Hylder and Judas Tree being among a couple of dozen common English names. Old tradition has it that Judas hanged himself upon an Elder, hence that name derivation.

1483 – In the late 15th century English plague text *Regimen Contra Pestilentiam* elder finds a place in causing pestilential buboes and swellings to ripen, as the text instructs thus: "*And the foner a fwellyng maye be made rype / take this medicyne as foloweth. Brufe the leuys of an Elder tree & putte therto ground muftard & make a playfter therof and putte it upon the fwellynge...*" [Regimen Contra Pestilentiam]

1596 – The *Rich Store-House or Treasury for the Diseased* recommends the following recipe for a '*Preferuative againft the Plague*' and contains what appears to be the leaves of that other well-known country plant, the bramble. It is not quite clear from the following whether the author is referring to a red berried form of bramble, or simply older, reddish coloured, leaves: "*Take a handfull of Hearbe-grace, otherwife called Rue, a handfull of Elder-leaues, a handfull of red Sage, and a handfull of red Bramble leaues, and ftamp them well together, and ftraine them through a fine linnen cloth, with a quarte of white Wine, then take a quantity of Cafe Ginger, and mingle it with them, and drinke a good draught thereof both morning and evening for the fpace of nine days as together, and by Gods grace it will preferve you.*"

In another remedy for a 'Medicine to breake the Botch' there is again mention of red bramble leaves, and elder: *"If it fortune the Botch to appeare, then take red Bramble leaues, Elder leaues, and Muſtard feede, and ſtamp them all together, and then take therof, and make a Plaiſter and lay it to the fore, and it will draw forth all the venome."* [Rich Store-House or Treasury for the Diseased]

1653 – Red bramble leaves appear again in the following recipe for a personal *'Defenſive'* against the plague, in addition to elder and several other other familiar herbal greens: *"Take Rew, Elder leaves, Sinkfoyle, or Tormentill, red Sage, red Bramble leaves, Sorrell, Marigold leaves, and Angelica ana j M. ſtamp them all in a Morter then put to them white Wine one quart, Wine vinegar d j. pinte, white Ginger powdered iiij 3, let it ſtand in a Pot cloſe ſtopped twenty foure houres, then ſtraine it hard forth; let the elder people take every morning two ſpoonfulls, faſt two houres after, and children one ſpoonful, uſe it nine mornings."* [Epitomie of Most Experienced]

1665 – The 1720 re-publication of Nathaniel Hodges' medical experiences in the Great Plague of London recommends that in treating the plague: *"Some Berries are alſo of great Uſe in Practice; as the Powder of* ivy Berries *given to the Quantity of one Dram in two Parts of Elder Vinegar, and one Part of* White-Wine; *the Spirit likewiſe drawn from* Elder Berries *would do the fame in a Doſe of* ℥ iij. *or* ℥ iv. *the Spirit of* Juniper Berries *given* ℥ i. *a Spirit drawn from green* Walnuts, *with* Treacle-Water, *as alſo from the Seeds of* Carduus, Citrons, &c. *had likewiſe their due Recommendations in powerfully promoting Sweat."* [Loimologia]

1668 – In Gervaise Markham's *English Housewife*, which can only be described as *the* domestic home-advice and home-help book of the 17th century, there is the following recipe for a 'Preſervation againſt the peſtilence' which is very similar to some of the previous ones above but combines ginger with treacle in the mixture: *"Take of* Sage, Rue, Brier leaves, *or* Elder leaves *of each an handfull, ſtamp them and ſtrain them with a quart of white wine, and put thereto a little Ginger, and a good ſpoonfull of the beſt* Treacle, *and drink thereof morning and evening."* [English Housewife]

1671 – In a domestic, or at least lay-medical, French book which draws its content from English medicine of the period there is an interesting reference in the section on plague remedies to one used by the King of England, *Remede du Roy d'Angleterre contre la Peſte*. Sage and elder are the leading ingredients with what is possibly raspberry (*Rubus idaeus*) leaves and possibly typographically incorrect in the original: *"Prenez Sauges, feüilles de Sureau, feüilles de Rubus idens [sic], de chacune demy poignée: Rhuë, Romarin, Aceta Oſella, de chacun demy poignée. Piles tout enſemblé dans un mortier, & le detrempez avec une pinte de bon vïnaigre de vin blan [sic], & une pinte de vin blan [sic], puis le paſſez dans un linge, & y ajoûtez un demy ſeptier d'eau Angélique. Faîtes diſſoudre dans cette liqueur une dragme de Mitridat, une dragme de Theriaque ou d'Orvietan.*

Prenez de cette eau une cuilerée, matin & foir, & ferez prefervé infailliblement." [Recueil des Remedes et Secrets]

1677 – The original Latin version of *Anatomia Sambuci, or the Anatomy of Elder* was published in Leipzig during 1631, with English editions appearing in 1650, 1655, and possibly 1670 too. The edition from which the following extracts are taken, falls under the section entitled '*Of the Pefte & peftilential fevers*', and does not appear to have been significantly altered from the original:

"*In curing and preferving from the Plague, great is the ufe of the Elder. A little fponge being wet in Vinegar of the Elder, and carried in a hollow globe made of Juniper wood, and fmell it, it mightily ftrengtheneth the fpirits, againft the impreffion of the infectious contagion.*

Red hot bricks, being befprinkled with this Vinegar, and a vapour roufed, it doth difipate the contagious virulency, fo that it cannot infinuate itfelf in mens houfes and cloths....

Rob of Elder and the extract prepared from it, here are excellent: the firft whereof is named by many, The Country-mans Theriock [ED – '*Theriaca Rusticorum*' in the 1631 Latin edition], *of which each week to fwallow the bignefs of a Walnut, and drink above it its proper Vinegar, and fo to fweat in bed, is a commonly received prefervative. This may be fitly ufed by thofe who are infected with the plague, efpecially if you mix it with fome of the anti-peftilential powders; or at leaft drink above it three or four fpoonfulls of Antilemick Vinegar of the Elder.*

The fame Rob chiefly it that is moft recent, being fpread more thickly on a fhive of bread, and eaten an hour or two before your meat, loofneth the belly; in whofe place you may give a fpoonful or two of the fyrup of the juice of the berries.

It is enough to fwallow fometimes in a morning before you go out the greatness of a peafe of the extract.

Rob, and the Extract Antilemick of the Elder

R. Roots of Tormentillae,
    Buterdock
    Of Pimpanels,
    Of Angelica,
    Leaves of Scordium,
    Berries of Juniper, of each
    half an ounce.

Macerate the roots 24 hours in Elder vinegar, afterwards dry them at leafure, and being powdered by themfelves, add the leaves of Scordium and berries of Juniper, likewife in powder; mix them all together, and with the Vinegar that remained befprinkle them, and work them moft exactly with a pound of Rob Sambuci, in form of an opiat: Of which give to the infected perfon two drachms in a convenient liquor, to provide fweat, and thruft out the poyfon from his heart.

The fpirit of the Elder by it felf is here very powerful, both in preferving, a few drops thereof being taken with a little white bread in a morning, and likewife in the beginning of the difeafe, a fpoonful or two being taken thereof before the feverish heat be powerful.

... Neither is it unwholfom, if once or twice a week in the morning, an hour or two before dinner, a cup of the wine prepar'd of the berries be taken but remember to take before it a little broth; for it loofneth the belly, hindreth the putrefaction, and by reafon of the Bezoartick vertue of the berries, it preferveth the body from contagion."

There is also a long recipe for a topical oil mixture that is made by leaving the ingredients in sunlight to infuse for a long time; a process that has similarities in the production method as for making St. John's Wort oil in many old herbal texts. For those that could not wait or were in a hurry *Anatomia Sambuci* has this quick-fix alternative: "*Some prepare it fuddenly thus, they take the oyl of infufed Elder-flowers, as much as is neceffary, in it they imerge the flowers of Marfh-Mallows and Hypericon, and boil them together in Bal. Mar. for fome hours; afterwards they exprefs ftrongly the flowers, and ftrain it; in the ftrained oyl they immerge recent flowers, boil them, prefs them, and ftrain them, and afterwards add Nitre.*

The way of ufing it is this; The whole body of the infected perfon within 24 hours is to be anointed with this oyl warm, and being wrapt in warm fheets, he is to be laid in a warmed bed to fweat; for they affirm that it is proved, that by this only remedy many have fafely efcaped the fiercenefs of this poifon..."

Another piece of advice for treating plague sores with elder is: "*Apply to Buboes peftilential, and Carbuncles, a Plafter made of the meat of Elder-flowers and Hony, which is excellent in ripening thefe tumours...*" [Anatomia Sambuci]

1722 – A variation on the elder-bubo remedy above is to be found in Scarborough's plague treatise. Under the section '*For* Buboes *and* Carbuncles' there is this recipe which adds mustard to the mixture: "*Rx. Elder-Leaves, red Bramble-Leaves, alike Quantity; which, with Muftard-feed, reduced into a Pafte, for a Plaifter to lay upon the Sore, to both draw and heal.*" [Practical Method as Used for the Cure of the Plague]

ELECAMPANE – *Inula helenium*

Stately elecampane, with its solitary golden-yellow flowers set on top of tall stalks, has a long history as a herbal plant. Plantlore suggests all sorts of name derivations for this species, including connections with Helen of Troy, and it was known on the Isle of Wight as the Wild Sunflower.

Perennial elecampane has a massive, succulent, rhizome that makes it reasonably drought tolerant, although the plant prefers moist pastures, meadows, and field borders, and was a regular in herbal gardens. According to some old accounts the Romans used the large roots, which have a camphor-like aroma, as an edible vegetable, while later herbalists esteemed elecampane for its tonic, expectorant and pectoral qualities. Indeed, candied elecampane lozenges were a popular cough cure, and French herbalists prepared the root in a liquid form known as *Vin d'Aulnee* for the same curative purposes.

In the medical treatise *Theory and Practice of Chirurgical Pharmacy* [1761] we are told also that the roots are: *"... gently irritating, and were formerly supposed to have a specific quality in curing the itch; and also to avail against the gout."* Enula roots were sometimes combined with other ingredients, in the form of ointments and liniments, one of which went by the name of *ungentum enulatum*. The 'itch', incidentally, is scabies.

Reaching up to five feet in height, and flowering during the Summer months, elecampane has quite large, lance-shaped, leaves that are velvety underneath, while the base leaves are long-stalked. In some parts of Britain the large broad leaves gave rise to the quaint common name of Elf Dock.

1543 – A mid-16th century French home recipe-remedy book (containing everything from whitening teeth to other household tips) has the following expensive plague potion, identified as a '*Remede treſbon quant tu yras en lieu ſuſpect de peſte*', which contains elecampane: *"Prens perles fine pilees, coral fin ambre grys, & muſc, de chaſcun cinq graīs demye once de clou batu auec racine de campane: faiz vng ſachet de ſandal cramoyſy, & metz les choſes ſuſdictes pulueriſees dedās, & te les applique ſur leſtomach, cela te gardera tresbien."* [Nouveau Traicte]

1544 – Thiebault's small fifty-something page French treatise on remedies for the plague, fevers, and some other ailments, has the following snippet for elecampane as a *preservatif* against pestilence. Chewing elecampane root would presumably provide bronchial relief in plague fever, and similar elecampane chewing remedies can be found elsewhere among early herbal texts: *"Et tiendrez en voſtre bouche une petite piece de zeduar / ou de la racine de Enula campana laquelle ayt trempe en fort vin aigre / par leſpace de vingt quatre heures.* [La Tresor du Remede Preservatif]

1575 – From the 1649 English version of Ambroise Paré's 16ᵗʰ century French medical and surgical publication we are told: "*Som* [sic] *think themfelvs fufficiently defended with a root of Elecampane, Zedoarie, or Angelica, rowled in their mouth, or chawed between their teeth.*"

Elsewhere in the text there is a quite interesting use of elecampane – as a body wash against the plague. As far as this writer can see this particular curative usage does not appear to have been a widely adopted remedy in the history of herbal plague treatments: "Let them wafh their whole bodies with the following lotion. Rx. *aquae rof. aceti rofati, aut fambucini, vini albi aut malvatici, an.* ℔ *.vi. rad. enulae camp. angelicae, gentia. biftorte, zeodar. an.* ℥iii. *baccar. juniperi, & hederae, an.* ℥ii. *falviae, rorifmar. abfinth. rutae, an.* m.i. *corticis citri,* ℥fs *theriacae & mithridat. an.* ℥i *conquaffanda conquaffent, builliant lento igni, & ferventur ad ufam antè commemoratum.*" [Workes of Ambrose Parey]

1585 – Castore Durante's 16ᵗʰ century herbal, published in Rome, flags up all the usual applications of elecampane for lung, asthmatic and similar complaints. The general rehearsed wisdom of writers such as Dodoens and Gerard was that the roots were better, or more potent, when dry, yet Durante's reference suggests green roots being used in relation to plague: "*La radice verde confettata, come il zenzero vale à i mali fopradetti, & gioua alla pefte, & al morfo de i ferpenti... La radice uer de gioua impiaftrata à i morfi de i ferpenti, alla pefte, & à tutte le pofteme peftifere.*" [Herbario Nuovo]

1636 – In *Haven of Health*, a work aimed at a lay-medical level, Thomas Coghan tells his readers: "*The common people, faith Hollerius, ufe to fteepe Elicampane rootes in Vineger* [sic], *and to lap them in a linnen cloth, and to carry them about with them, fmelling to them often times. Others before they goe forth in the morning eate Garlike, and drinke a draught of new Ale after it, or good Wine. But Garlike is thought of many to bee rather hurtfull than wholefome in the Plague, becaufe it openeth the pores of the body too much, and fo maketh it more apt to receive infection.*" [Haven of Health]

1667 – The Company of Distillers of London had gained a virtual monopoly over the production of spirits and strong waters since the grant of their Charter in the 1630's, and one of the products that they produced was a type of plague water called '*Pretious water*' which is described in their publication *The London Distiller* as: "*Pretious water is good againft the Plague and Malignant Feavers: It alfo comforteth the Spirit, ftrengtheneth the Heart, preferveth the Senfes, and relieveth languifhing Nature.*" Two recipes are given, for large and small quantities, and the one below produces the greater quantity. From a plant interest perspective it would appear the roots of wood avens (*Geum urbanum*), which have the smell and taste of cloves, and sweet chervil, otherwise known as sweet cicely, have places in the ingredient list:

"Take ftrong Proof-fpirit 10 gallons, the roots of Enula Campana, Cyprus, Avens, Calamus Aromaticus, Angelica, Saffafras of each 5 ounces: Zedoary, Galingale, of each four ounces, Caffia lignea, Lignum Rhodium, Yellow Saunders of each 3 ounces, Citron pils dry, Orange pils dry, of each 6 ounces, Cinnamon White, Nutmegs, Mace, Ginger, of each five ounces, Cinnamon beft 20 ounces, Cloves, Cardamums, Cubebs, of each 2 ounces and a half, fweet Chervile feeds, Bafil feeds, of each 3 ounces and half, Coriander feeds, fweet Fennel feeds of each ten ounces, Annifeeds 20 ounces; bruife them, diftil them into Proof-fpirit, and dulcifie with fine Sugar 15 pound according to Art: Let it ftand till it be fine, then draw it off, and add Musk one dram: Ambergrife 4 drams; then let it clear and draw it for ufe." [The London Distiller]

1747 – James' somewhat expansive and professionally-orientated Pharmacopoeia Universalis has the following regarding elecampane: "The Root is the Part us'd, which is both pulmonic and ftomachic, alexipharmic and fudorific. It is chiefly us'd in Coughs, Afthmas, Crudities of the Stomach, in opening the urinary Ducts; in the Plague, and other contagious Diftempers. Externally it is recommended in the Itch, Spafms, and Ifchiadic Pains." [Pharmacopoeia Universalis]

1747 – Short's domestic Medicina Britannica gives most of the usual, traditional, pectoral qualities of elecampane, but in relation to plague says: "... eaten candied like Ginger, it is good againft the Plague." [Medicina Britannica]

1751 – The History of the Materia Medica by John Hill was aimed at professional medics, and gives us a little insight into the apothecaries' 'trade' in passing, though it is not specific regarding the plague and elecampane use: "Wine in which the Roots of Elecampane have been infufed is efteemed a great Prefervative againft Contagion... it is an ingredient in many of the Compofitions of the Shops..." [History of the Materia Medica]

1770 – Alston's Lectures on the Materia Medica doesn't make the root sound overly pleasant to the palate. Curiously, it looks as though Alston has no practical know-ledge of the plant's properties (referring to the spicy taste lasting 'above an hour'): "The tafte is at firft difagreeable, as it were rancid, like foap; then aromatic and bitter, and at laft very hot; the heat, though not painful, continuing I think above an hour: the fcent alfo is pretty ftrong, partly aromatic and partly fetid." [Lectures on the Materia Medica]

1852 – Almost one hundred years later and the comments by Thompson in the eleventh edition of the London Dispensatory would appear to suggest that elecam-pane had fallen into disuse: "It was formerly regarded as a remedy of great efficacy in dyspeptic affections, flatulencies, palsy, dropsies, uterine obstructions, and pulmonary complaints; but Cullen observed, that its diuretic powers were trifling; and could not discover that it possessed any expectorant properties. It is scarcely ever used by the regular practitioner." [London Dispensatory, 11th Ed.]

CLOVE-GILLYFLOWER – *Caryophyllus* sp.

The clove gillyflower, or gilliflower, is related to carnation, and pinks also, with an almost bewildering number of varieties in each category. Both carnations and clove-gillyflowers were one of the chief ornamental flowers in English gardens during the 16th and 17th centuries, but it is the gillyflowers that most interested herbalists.

In *Paradisus Terrestris* [1629] the apothecary-botanist John Parkinson says that the way his contemporaries distinguished between the two species, in terms of scientific naming, was to call carnations *Caryophyllum sativus maximus*, and gillyflowers *Caryophyllum sativus maior*. The red, or clove gillyflower, he says: *"... is most vsed in Physicke in our Apothecaries shops, none of the other being accepted or vsed (and yet I doubt not, but all of them might serve, and to good purpose, although not to give so gallant a tincture to a Syrupe as the ordinary red will doe) and is accounted to be very Cordiall."* Parkinson identifies *Caryophyllus ruber* as the specific plant to be used.

Not all sources were convinced that clove gillyflowers had the slightest use whatsoever as appears, for example, when reading between the lines of the entry for gillyflowers in *The New Dispensatory* [1753]: *"Clove july flowers. A great variety of these flowers are met with in our gardens: those made use of in medicine ought to be of a deep crimson colour, and a pleasant aromatic smell, somewhat like that of cloves; many sorts have scarce any smell at all. The caryophylla rubra are said to be cardiac and alexipharmac [sic]: Simon Paulli relates, that he has cured many malignant fevers by the use of a decoction of them; which he says powerfully promotes sweat and urine, without greatly irritating nature, and also raises the spirits, and quenches thirst. At present these flowers are chiefly valued for their pleasant flavour, which is entirely lost even by light coction: hence the college direct the syrup, which it the only officinal preparation of them, to be made by infusion."*

Writing in the early 19th century John Waller [1822] also refers to clove gilly-flowers in the past tense, although he does mention a decoction or infusion of the plant being useful in typhus fevers: *"These flowers are in esteem as possessing considerable cordial and cephalic properties; it was justly valued by the ancient medical writers as a remedy of some value in low putrid fevers and nervous head-aches. The preparations from this plant, formerly kept in the shops, were a syrup, conserve, distilled water, and a preparation with vinegar."*

1578 – Certainly during the 16th century the herbal qualities of gillyflowers had not been questioned in terms of their use in treating the plague as the Lyte-Dodoens text suggests: *"The Conserve of the floures of the first kinde, made with Sugar, comforteth the harte, & the use thereof is good against the hoate feuers & the Pestilence."* [New Herbal]

1579 – In William Bullein's *Bulwarke of Defence* gillyflower roots, rather than the flowers, are promoted as useful: "*The roote is good agaynſt the Peſtilence and falling ſickneſſe...*"

1629 – John Parkinson's gardening book, *Paradisus Terrestris*, does not have proper herbal remedies but recommended the following method for conserving gilly-flowers that could have been commonly used in remedies: "*Cloue Gilloflowers... preſerued or pickled vp in the... manner (which is* ſtratum ſuper ſtratum, *a lay of flowers, and then ſtrawed ouer with fine dry and poudered Sugar, and ſo lay after lay ſtrawed ouer, vntill the pot bee full you meane to keepe them in, and after filled vp or couered ouer with vinegar)...*" [Paradisus Terrestris]

1665 – Under the section about carnations in Lovell's *Panbotanologia* we find: "*Vertue. the conſerve of the flowers of the clove-gilliflower is very cordiall and exhilerating, it helpeth hot peſtilentiall feavers, and expelleth the poyſon, and fury of the diſease.*" [Panbotanologia]

1707 – The French de Ville publication of Caspar Bauhin's work similarly reflects use of the flowers; the first de Ville reprints of Bauhin's original appearing in the late 17th century: "*... la conſerve qu'on en fait ſert contre le venin de la peſte; chaſſe les vers qui ſont dans le corps, l'eau diſtilée de toute la plante & ſur tout des feüilles a les mêmes effets...*" [Histoire des Plantes de L'Europe]

1710 – William Salmon, in *Botanologia*, differentiates between clove-gillyflowers and carnations, the flowers of gillyflower being much smaller than those of carnation. Among the uses Salmon recommends:

"The Inspiſſate Juice. *It is highly cordial, and may be diſſolved in white Port Wine or Canary, to make a Tincture of at pleaſure, againſt fainting and ſwooning Fits, Sickneſs at Heart, Malignity of the Plague and Poiſon.*

The Syrup made with Water, *as is taught in our* Pharmacopoeia Londinensis... *Take Clove-gilliflowers (the Whites being cut off) a pound: infuſe them all night in Spring Water two pounds: being ſtrained, with double-refined white Sugar four pounds, make a Syrup (without boiling) only by melting the Sugar. This Syrup is temperate in Quality, ſtrengthens the Heart, Stomach and Liver; it is Pectoral and Cordial, and maybe mixt with other Cordial Liquors, againſt malignant Fevers, and the Plague...*

The Spiritous Tincture. *It is a great Preſervative againſt all manner of Malign, Infectious and Peſtilential Diſeases, and ought to be uſed preventively, as firſt in the Morning faſting; ſecondly about four in the Afternoon; thirdly at Bed time. It more powerfully comforts the Stomach, chears* [sic] *the Heart, and revives the Spirits, than any of the former preparations. It may be prepared with common Spirit of Wine: and then it may be take alone of it ſelf, from one ſpoonful to two, according to the Age of the perſon; or otherwiſe mixt with Wine, or ſome other Vehicle.*" [Botanologia]

1713 – In Salmon's re-published version of Bate's *Pharmacopœia*, and which is aimed at medical professionals, he identifies a cordial water where gilly-flowers are teamed up with dates, as well as some other exotic aromatic ingredients:

"*Aqua Caryophyllorum*, Water of Clove-Gillyflowers.
Bate.] Rx *Clove-Gillyflowers new gathered* ℔ij. *Dates*, ℔j. *Annifeeds, Liquorice, ana* ʒj. *Lees of Wine* ℔xxiv. *digeſt and diſtil in an Alembick* ℔iv. *to which add Sugar-candy* ʒiv. *and hang it in a Nodule* ; *Cloves* N° iv. *Muſk and Ambergrieſe ana gr.* v.

Salmon.] It is an excellent Cordial-water, good againſt Faintings, Swoonings, Sickneſs at Heart, Opreſſion of the Spirits, Wind, Gripings, and fuch-like; and is very profitable to be given againſt the Plague, and all Manner of peſtilential Fevers, proceeding from Putrefaction." [Pharmacopœia Bateana]

1722 – A hundred years on from Parkinson's gillyflower conserve of the flowers Quincy's words seem to suggest that the conserve was falling out of favour: "*They are a fine* Aromatick, *and very grateful to the* Smell *and* Taſte. *They have place in the Syrup made of them, and in moſt Cephalick and Cordial* Juleps. *There is alſo a* Conſerve *made of them, but hardly ever uſed.*" [Pharmacopœia Officinalis]

1751 – In the *History of the Materia Medica* Hill advises readers that only the syrup of *Clove July Flowers* or *caryophylli flores*, should be regarded as useful, with other gillyflower preparations having little value: "*Theſe Flowers have great Commendations given them by the medical Writers as Cordials. They are recommended in all the Diſorders of the Head, in Palpitations of the Heart, and in nervous Complaints of whatever Kind: They have been alſo much praiſed in malignant and peſtilential Fevers.* Simon Paulli *tells us, with an Air of great Certainty and Aſſurance, that he had cur'd great numbers of People of malignant and peſtilential Fevers, by no other Medicine but a ſtrong Infuſion of theſe Flowers in Water, which he tells us is a powerful Sudorific and Diuretic, and that it at the ſame Time comforts the Patient inſtead of weakening him. There uſed to be a diſtill'd Water of theſe Flowers kept in the Shops, but it has been found to poſſeſs very little of the Virtue it was once ſuppoſed to have, and has accordingly been rejected: The only Preparation now in Uſe is the Syrup made from a ſtrong Infuſion of the Flowers in Water; this is a very fragrant, well taſted, and high colour'd Syrup, and is much uſed for the ſweetening Juleps, and other of the common liquid Forms of Medicines, but is not depended much upon as a Medicine itſelf. A Conſerve is made by ſome of the Flowers, and is a very pleaſant one and keeps well. It ſerves excellently to bring the Cordial Electuaries into Form, but it is not univerſally in Uſe. Some alſo macerate the Flowers in clear and ſtrong Vinegar, they give this Liquor a high Colour, a fine Flavour, and very fragrant Smell; this is call'd* Acetum Caryophyllatum. *It is eſteem'd excellent againſt peſtilential Contagion. People are adviſed to take a Spoonful or two of it before going into a ſuſpected Place, and always to carry in Times of Infection a Handkerchief wetted with it in their Pockets, that it may be ready to be ſmelt to on Occaſion; but in this Caſe it is but Juſtice to obſerve, that the Vinegar may have a greater Share in the good Effect than the Cloves.*" [History of the Materia Medica]

JUNIPER – *Juniperus communis*

Slow growing juniper is an evergreen shrub or small tree, very often reaching no more than five to seven feet in height, and prefers calcerous soils and dry hilly or heath-like habitats across Europe. The leaves have a slight blue-green colour and bristle-like structure, while the whitish-yellow flowers give rise to purplish-black aromatic berries with a pungent sweet taste that find their way into cookery recipes and, in the past, traditional medicinal recipes including those to treat the plague. Juniper wood was also among the aromatics recommended to be used as fumigants to cleanse the air in streets and private homes.

The berries were used in carminative and stomachic preparations, the herbal qualities of the berries being extracted in the form of a spirituous water or distilled essential oil, and were commonly kept in apothecaries' shops. One particularly valued preparation from juniper berries was a 'rob' that was prepared from the dregs of berries used to make the essential oil. The virtues and production of the rob are described in *The New Dispensatory* [1753] as follows: *"The liquor remaining after the diſtillation of the oil, paſſed through a ſtrainer, and gently exhaled to the conſiſtence of a rob, proves likewiſe a medicine of great utility, and in many caſes is perhaps preferable to the oil, or berry itſelf: Hoffman is expreſsly of this opinion, and ſtrongly recommends it in debility of the ſtomach and inteſtines, and ſays it is particularly of ſervice to old people who are ſubject to theſe diſorders, or labour under a difficulty with regard to the urinary excretion: this rob is of a dark, browniſh-yellow colour, a balſamic ſweet taſte, with a little of the bitter, more or leſs according as the ſeeds in the berry have been more or leſs bruiſed."*

1575 – In the first English edition of Ambroise Paré's French-language medical and surgical publication, which first appeared in 1575, the translated wisdom on juniper in a cordial water (used as both a 'preservative' against plague but also to treat those who were already infected), was: *"This Cordial water that followeth if of great vertue. Take of the roots of the long round* Ariſtalochia [sic], *Tormentil, Diptam, of each three drams, of Zedoarie two drams,* Lignum Aloës, *yellow Sanders, of each one dram, of the leavs [sic] of Scordium, St. John's-wurt, Sorrel,* Rue, Sage, *of each half an ounce, of* Bay *and Juniper-berries, of each three drams, Citron ſeeds one dram,* Cloves, Mace, Nutmegs, *of each two drams, of Maſtich, Olibanum, Bole-Armenick,* Terra Sigillate, *ſhaveings of Hartſ-horn and Ivorie, of each one ounce, of Saffron one ſcruple, and the Conſervs of Roſes, Bugloſs-flowers, water-lillies and old Treacle, of each one ounce, of Camphire half a dram, of aqua vitae half a pinte, of white wine two pintes and a half, make thereof a diſtillation in* Balneo Mariae,[sic] *The uſe of this diſtilled water is even as Treacle water is."* [Workes of Ambroise Parey]

1578 – The Dodoens-Lyte herbal says: *"Juniper or the beries thereof burned, driueth away all venemous beaſtes and all infection and corruption of the ayre: wherefore it is good to be burned in a plague time, in fuche places where as the ayre is infected."* [New Herbal]

1579 – From the same time period Bullein's *Bulwarke of Defence* adds in head colds to the 'uses' list: *"Jenuper beries be holfome to clenfe ẙ Stomacke, help the coughe, inflations, and torments of the belly. There is a precious Oyle made of Jenuper to warme the fynewes and comfort the head, being ouercome with colde. Jenuper beries be holfome to put in medicine agaynft the Peftilence, & byting of Serpents."* [Bulwarke of Defence]

1665 – A pamphlet issued by the Royal College of Physicians of London during the Great Plague suggests consumption of juniper berries as almost medicinal cuisine: *"Steep Juniper-berries in Vineger for a night, let the Vineger be exhaled off; eat thereof at pleafure."* [Certain Necessary Directions]

1665 – From the same time period Lovell describes uses of juniper oil: *"Oleum baccarum Juniperi Chym. Pharm. Lond. or the* Chymicall oile of juniper berries, *is excellent in pains of the yard, gonorrhæa and epilepfie. It's a great prefervation againft the peftilence and all evill aire, and helpeth the dropfie..."* [Panbotanologia]

1666 – Boghurst, who gives one of the best medical accounts of the Great Plague of 1665, provides alternatives to the commonly used Ruffi pills (page 194) – the *Pilule Aloephanginae*, and then the following extract:

"Rx. rad. angelicæ, imperatoriæ, enulæ, gentiani, Zedoarie, baccar. Juniperi, rad. serpentariæ, vel gentianæ ana ℥ s.s. fol. scordii, cardui, rutæ, ana M. i. Cinnamoni, charyophyllorum, croci, dictamni cret. ana 3 ii Cort. extern. Citri, summitat. Centaur, ana ℥i folior. sennæ, agarici ana ℥i s.s. rhabarbari 3 i epithymi ℥ s.s. infundent omnia contusa in vini albi q.s. per septimanam deinde coquentur leviter et colaturæ claræ adde aloes optim. ℥vi Myrrhæ claræ rub ℥i s.s. balsami peruv., Mastickes, ol. nuc. Mosch. per express. ana 3 ii M. et evaporando redigatur in Massam idoneam. Of this pill may bee taken two scruples once in 4 or 5 dayes, but as for myself, I never took a grain of any pill or any purging physick these many years, though I count not this for a rule for everybody..." [Loimographia]

1703 – The early 18th century English translation of Etmuller's German work *Practice of Physic* put the rob of juniper to use in cases of plague as follows: *"Take of the Rob of Juniper, three ounces; Diafcordium and Mithridate, of each an ounce and a half; Venice Treacle, half an ounce; Flowers of Sulphur, two or three ounces; Myrrh, three drams; Frankincenfe, fix drams; Camphyr, two drams; Saffron, half a dram; Juice of Citrons, half an ounce. Make an Electuary. Dofe a dram and a half in a convenient Vehicle."* [Etmullerus Abridg'd]

1713 – Salmon's professional readership work *Pharmacopœia Bateana* offers up a recipe for a juniper vinegar to be used as a gargle or fume (one presumes a

161

fumigant and not inhaled steam, given that plague victims on death's door would have been capable of little personal activity like inhaling steam over a bowl):

"*Acetum Peftilentiale*, Vinegar againſt the Plague.
Bate.] Rx *Roots of Angelica, Zedoary*, A. ʒj. *Juniper-berries* ʒij. *Rue*, M. iij. *beſt Wine-vinegar* ℔iij. *mix, digeſt, and ſtrain.* It is uſed for a Fume; and to waſh and gargle the Mouths and Throats of ſuch as are affaulted with the Plague.
Salmon.] If *Camphire* ʒij. *fine Pouder of Bay-berries, and Winters Cinnamon*, A. ʒſs. *Spirit of Wine* ʒviij. *be added*; the Medicine will be much more ſpirituous and effectual to the Purpoſes intended." [Pharmacopœia Bateana]

1722 – From John Quincy's text it would appear that juniper was not always seen as useful. Regarding the berries: "*Some will have them to be great Strengtheners of the* Stomach, *and effectual againſt* Malignities *and* Epidemical Infections. The* Rob *(a Form of Medicine now out of uſe amongſt us) made of their expreſſed Juice, when green, is call'd the Theriaca Germanorum, ſo much were they in eſteem amongſt them for their* anti-peſtilential *Qualities*." [Pharmacopœia Officinalis]

1751 – John Hill's comment on the spiritous water of juniper, and its migration into domestic gin production, shows an interesting social sideline to the avail-ability of juniper spirit among the apothecaries' herbal remedy output: "*The* Germans *think them the greateſt of all Medicines againſt Contagion, and hold it almoſt impoſſible to receive Infection, even the Plague, while they have ſome of the Berries in their Mouth... We uſed to keep a diſtill'd ſpiritous Water of Juniper in the Shops, but the vulgar got an Opinion of its being a pleaſant Dram, and conſequently the making it became the Buſineſs, not of the Apothecary but of the Diſtiller, who ſold it under the Name of* Geneva." [History of the Materia Medica]

1770 – While Alston refers to the wood of juniper being burnt to purify contagious air his long list of juniper's herbal uses does not specifically suggest juniper to be used in anti-plague remedies. However, the very nature of some of properties listed would, presumably, have been recognized immediately by any physician reading this work which was aimed at medical professionals: "*The fruit is an antiſeptic, attenuant and ſomewhat deterſive, diaphoretic, diuretic, and carminative; called cephalic, pectoral, ſtomachic, uterine, alexipharmic, lithontriptic, &c. and commended in coughs, conſumptions, flatulent and nephritic colics, gravel, dropſy, menſibus obſtructis, hyſteric fits, ſcurvy, ſcabies, &c. The lignum is ſaid to be as good a ſudorific in the lues Gallica as the guaiacum; and to purify contagious air, if burnt. The vernix is made diuretic and anodyne; and diſſolved in oleo lini is a good ointment for the hemorrhoids, burnings, &c.*" [Lectures on the Materia Medica]

MARIGOLD – Calendula officinalis

The well-known garden marigold was a frequent ingredient in many remedies, no less so in the case of plague where the flowers were the focus of attention rather than plant foliage. The updated *New Dispensatory* [1753] refers to marigold as: "*... common in gardens, where it is found in flower* [the] *greateſt part of the Summer. Marigold flowers are ſuppoſed to be aperient and attenuating; as alſo cardiac, alexipharmac, and ſudorific: they are principally celebrated in uterine obſtructions, the jaundice, and for throwing out the ſmall pox. Their ſenſible qualities give no foundation for any of theſe virtues: they have no taſte and very little ſmell. The leaves diſcover a viſcid ſweetiſhneſs accompanied with a more durable ſaponaceous pungency and warmth: theſe ſeem capable of anſwering ſome uſeful purpoſes, as a ſtimulating, aperient, antiſcorbutic medicine.*" The above précis seems to suggest that by the mid-eighteenth century medics were not overly convinced about the properties of marigold despite what previous generations had believed. Indeed, not all of the early works quoted here even mention marigold in relation to plague. In *Bulwarke of Defence* [1579] Bullein merely mentions of marigold that: "*... this herbe ſodde in wyne, drunke warme, moueth ſweat...*"

By the time of Green's *Universal Herbal* [1823] marigold had lost much of its herbal glory: "*It has however been cultivated time out of mind in kitchen gardens for the flowers, which were dried in order to be boiled in broth; from a fancy that they are comforters of the heart and spirits. Linnaeus accordingly says, that they may be used in a double dose, as a succedaneum to Saffron; but modern practice has little confidence in these supposed cordials.*"

1629 – The London apothecary-botanist John Parkinson recommended that the readers of his gardening book should grow a number of key herbal plants to make them more self-sufficient where medical help was not close by. Marigolds were one of his recommended plants: "*The herbe and flowers are of great vſe with vs among other pot-herbes and the flowers eyther greene or dryed, are often vſed in poſſets, broths, and drinkes, as a comforter of the heart and ſpirits, and to expel any malignant or peſtilential quality, gathered neere thereunto. The Syrupe and Conſerue made of the freſh flowers, are vſed for the ſame purpoſes to good effect.*" [Paradisus Terrestris]

1665 – In *Loimologia* [1720], the reprint of Hodges' account of the London plague, the following alexiterial water appears under the heading '*A Compound Anti-peſtilential Decoction*' and contains marigold and rose flowers, and other herbal ingredients listed elsewhere in this book:

"Rx. *Radic. Contrayerva ʒ j. ſcorzoneræ hiſpan. angelicæ ana ʒ j. B. fol. Scordii gelagæ ana M. iij. fl. roſ. rubr. calendulæ ana p. iij. raſ. C. C. eboris and ʒ j. bacc. juniperi, hederæ ana ʒ ij. in aceto ſambucino per triduum macerat: dictamni cretici,*

163

cortic. limonum ana ʒ j. ʃucc. meliʃʃæ, cardui, angelicæ ana lib. ʃs. aceti opt. lib. ij. diʃtilla in organis humilorbius poʃt. deb. inʃuʃionem." [Loimologia]

1665 – Lovell, in *Panbotanologia*, comments: *"The* conferve of the flowers *and ʃugar taken faʃting in the morning help trembling of the heart, and prevents the plague, &c. by the corrupt aire."* [Panbotanologia, or Compleat Herbal]

1666 – Thomas Willis' notes on remedies for plague were published in 1691, after his death, but the following were possibly used by Willis around the time of the London Plague when the publisher says they were collated: *"In time of Sweating, give the Patient Poʃʃet Drink made with Peʃtilential* Vinegar; *boyl in the Milk* Scordium *or* Marigold *Flowers; if he is very dry, boyl* Medefweet, *or* Wood Sorrel; *if he is ill at Stomach, and apt to vomit or faint, give* Claret *Wine burnt with* Cinnamon *and* Zeodary *Root, and* Mint Water mix'd with it: *Or elʃe give him Beer boyl'd with a Cruʃt of* Bread *and* Mace, *and ʃweetn'd with* Sugar."

The key *Pestilential Vinegar* ingredient mentioned above was made as follows: *"Take of the Roots of* Angelica, Butter-Burr, Tormentil, Elecampane, *of each half an ounce,* Virginian – Snake-Weed, *choice* Zedoary, Contrayerva, *of each three Drams; Leaves of* Scordium, Rue, Goats-Rue, *of each one handful;* Marigold Flowers, Clovegilloflowers, *of each half a handful; Seeds of* Citron *and* Carduus, *of each two Drams; Cut and bruiʃe theʃe, and put them in a Glaʃs-Bottle, with three Pints of the beʃt* Vinegar, *to digeʃt for ten days."* [Plain & Easie Method]

1710 – Salmon's domestic orientated *Family Dictionary* mentions a water of marigold flowers but also an *'Excellent Conʃerve made of them'*, the recipe for which went as follows: *"Take of Marigold Flowers two ounces. Confection of Kermes and Hyacinth, of each two drams, the Pouder of Pearl an ounce, and as much Syrup of Citron as will make them into a Conʃerve, by mixing and bruiʃing them well together with a ʃufficient quantity of fine Sugar.* Take of it about a quarter of an ounce Morning and Evening. It is a great Cordial for refreʃhing the Spirits, and a Preʃervative againʃt the Plague and Peʃtilential Fever." [Family Dictionary]

1727 – Of marigold flowers Strother says: *"Their Preparations are,* First, *A Water.* Secondly, *A Vinegar, which is eminently cephalick, whether ʃnuffed up the Noʃtrils, or apply's outwardly. In the Plague it has been adminiʃtered, in order to promote Sweats.* Thirdly, A Conferve, which is good in the aforefaid Cafes." [Materia Medica]

1739 – Elizabeth Smith's *Compleat Housewife* was a truly domestic book of kitchen recipes, household tips and home remedies, and went through numerous editions during the 18th century. The book lists the following recipe for a 'plague-water' which would have been just as complicated to produce as many of the potions produced by practising apothecaries but also contains many of the other herbal plant species covered elsewhere here: *"Take roʃa ʃolis, agrimony, betony, ʃcabius,* centaury-tops, ſcordium, balm, rue, wormwood, mugwort, celandine, roſemary, marigold leaves, brown ſage, burnet, carduus, and dragons, of each a large handful; and angelica-roots, piony-roots, tormentil-roots, elecampane-roots and liquorice, of each one ounce; cut the herbs, and ſlice the roots, and put them all into an earthen pot, and put to them a gallon of white wine and a quart of brandy, and let them ſteep two days cloſe cover'd; then diſtil it in an ordinary ſtill with a gentle fire; you may ſweeten it, but not much." [Compleat Housewife]

1747 – Short quotes the following domestic marigold remedy from Bartholine in *Medicina Britannica*: "*For a Preſervative, drink every Morning a Spoonful of Tincture of Marygold Flowers drawn in ſtrong Vinegar.*" Elsewhere Short tells us that: "*The cut Flowers, eaten in a Sallad with Oil and Vinegar, are an Antidote againſt the Plague; and, for Cure, an Ounce or two of the Juice of the Flowers drank faſting, and Sweat after it, this powerfully expels the Poiſon of the Plague.*" [Medicina Britannica]

1762 – As a cure for the plague preacher John Wesley suggested drinking marigold flower juice in 1761, then broadened his scope to include regular consumption of marigold flowers in the 1762 edition, but both examples pick up on previous herbal knowledge: "*Eat* Marigold Flowers *daily as a Sallad, with* Oil and Vinegar... *Or, an Ounce or two of the Juice of* Marigolds." [Primitive Physick]

1790 – The description of marigold uses in William Meyrick's family herbal does not specifically suggest marigold as a plague remedy, but possibly there may have been enough common knowledge of the plant's properties to link its past uses in the plague with Meyrick's comments on use of the flowers in smallpox and measles: "*A water diſtilled from them is good for inflamed and ſore eyes. A decoction of the flowers in poſſet drink is much uſed among country people as an expulſive in the ſmall-pox and meaſles. The leaves of the plant, when chewed, at firſt communicate a viſcid ſweetneſs, which is followed by a ſharp penetrating taſte, very durable in the mouth, but not of the hot or aromatic kind, but rather of a ſaline nature. The expreſſed juice, which contains the greateſt part of this pungent matter, has been given in doſes of two or three ounces, with a view to looſen the belly, which it ſeldom fails of doing: and it likewiſe promotes the other ſecretions of the body in general.*" [New Family Herbal]

1802 – In France the author of *Le Medecin Herboriste* has a the following recipe for a '*Décoction contre la peste*' which perhaps suggests that French physicians in the early 19th century still viewed marigold as a viable anti-plague ingredient: "*Prenez racines de pétasite, une demie-once; feuilles d'ulmaire, de chardon-bénit et de chamœdris, de chacune demie-poignée; fleurs de calendule et de pavot rouge, de chacune une pincée: faites-les cuire dans trois chopines d'eau de fontaine pendant un quart-d'heure, pour boisson ordinaire.*" [Le Medecin Herboriste]

MASTERWORT – *Peucedanum ostruthium*

Masterwort, which was also known as *Imperatoria*, is a member of the carrot (*Umbellifer*) family, some other members of which are among the anti-plague plant species included here. Growing to about two or three feet in height, stout masterwort is naturally at home in somewhat moist conditions, and so is found in meadows, and also grassy mountain habitats.

Masterwort was cultivated in English gardens, though in the early 19th century Thornton [1814] suggests that: "... *the root so produced is greatly inferior to that growing in the south of Europe, especially in mountainous situations; hence the shops are supplied with it from the Alps and Pyrenees.*" That insight into the herbal home-remedy industry is curious since Thornton follows up his comment above by bemoaning the fact that the plant "... *is omitted from our pharmacopoeias,*" in reference to official medical practice. Clearly master-wort had gone out of favour, yet in the previous century master-wort root was viewed as a valuable alexipharmic and sudorific herbal ingredient, and was seen as: "... *serviceable in Distempers and Contusions, in phlegmatic disorders of the Head, Palsy, Apoplexy, and in Crudities of the Stomach, and the Colic.*" [Pharmacopœia Univeralis, 1747]

1578 – The Dodoens-Lyte herbal refers to masterwort as '*great pellitory of spayne*' among the common names, and perhaps relates to the imported variety which was possibly the preferred ingredient of apothecaries. In the plague sickness the plant appears to have had a variety of applications that it was useful for: "*Maſterworte is not onely good againſt al Poyſon, but alſo it is ſinguler agaynſt all corrupt and noughtie ayre, and infection of the Peſtilence, if it be dronke in wine and the ſame roote pounde up by itſelf or with his leaues, doth diſſolue and cure Peſtilential Carboncles and Botches, and ſuche other apoſtumations and ſwellinges, being applied therto.*" [New Herbal]

1585 – Durante states that the virtues were somewhat similar to those listed by Dodoens, masterwort being used in fever but also for plague sores: "*Di fuori. Applicata con teriaca, & aceto roſato al cuore nelle febri peſtilentiali, lo conforta mirabilmente. Il ſucco conſuma & leua via applicato le carne putride della cācrene. L'herba peſta mitiga il dolore della ſciatica, & riſolue i tumori.*" [Herbari Nuovo]

1665 – A similar case is made by Lovell in *Panbotanologia* nearly a hundred years later: "*Vertue. drunk with wine it helpeth againſt all poyſon, peſtilence, and corrupt aire. The roots and leaves ſtamped and applied help peſtilential botches, and ſuch like ſwellings.*" [Panbotanologia, or Compleat Herbal]

1665 – A pamphlet published by Thomas Cocke during the Great Plague of London, and entitled '*Advice for the Poor by way of Cure & Caution*' was much more ambitious with masterwort, suggesting the following '*Cordiall Antidote againſt the* Plague': "*Take* Sage, Rue, *of each one handfull,* Maſterwort root,

Butter-bur *root*, Angelica *roots, and* Zedoary *of each half an ounce*, Virginia Snake-*root a quarter of an ounce*, Safron [sic] *20 grains, Contra Yerva a dram* (at the Herb ſhops and Drugſters you may have them all) Malago Wine *a quart, bruiſe the herbs, and pound the roots, and put them in a Pipkin cloſe covered, and ſet it to the fire, and let it ſtand hot, but not boil, for the ſpace of an hour or better, then ſtrain it out, and put in a quarter of an ounce of* Mithredate, *and as much* Venice Treacle, *which diſſolve in it. Take hereof half a Spoonfull every Morning firſt, and every Night laſt, for preſervation; but if one be taken ſick, then let them drink quarter of a pint, and cover them to ſweat. This drink will powerfully fortifie the vitals, and by ſweat throw out the* Malignity *of the diſtemper.*" [Advice for the Poor]

1667 – In *The London Distiller*, a manual for members of The Company of Distillers of London which had a virtual monopoly over the production of spirits and strong waters since the 1630's, is the following recipe for their plague water that includes masterwort (imperatoria) among some other familiar herb ingredients:

"*Take ſtrong Proof ſpirit 1 gallon, Butter-bur-roots dry one ounce and 5 drams, Garden Valerian roots dry, Common Valerian roots dry, Angelica roots, Imperatoria, Gentian, Enula-Campana, Snake-graſs roots, of each half an ounce and three quarters of a dram, Contra Yerva, Zedoary, Galingale, of each 3 drams and a quarter, Rue leaves dry, White Horehound, Scordium, Carduus Benedictus, of each half an ounce, Elder flowers, Lavender, Mace, of each two drams and an half, Citron pils dry, Juniper berries, of each 6 drams and a half, Green Walnuts with the husks 1 ounce 5 drams, Venus* [sic] *Treacle, Mithridate, of each a dram and half, Anniſeeds beſt 2 ounces 3 drams and an half, Camphire 3 quarters of a dram; diſtil into ſtrong Proof-ſpirit according to Art; dulcifie with white Sugar what ſufficeth; for uſe; let the party infected take of this water one ounce mingled with warm Poſſet drink, (or any water proper in that caſe) and be kept very warm, and ſweat well thereon.*" [London Distiller]

1710 – In *Botanologia* William Salmon refers to masterwort root being chiefly used and lists the following ways of using the plant in anti-plague remedies:

"The Eſſence... *it has alſo a ſingular Virtue and Power againſt all ſorts of cold Poyſons, as alſo againſt any malign Diſeaſe, and the Plague itſelf: it provokes Sweat, and defends the Heart againſt any Venom, Malignity or Infect*ion.

The Infuſion or Decoction of the seed in Wine. *It has all the Virtues of the* Infuſion *or* Decoction *of the Root, and may be given in the ſame manner and ſame Doſe. It is a ſingular Prophylactick againſt peſtilential Airs, and againſt the very Infection of the Plague itſelf.*

The Acid Tincture of the Root or Seed. *It is a potent thing againſt the Plague, and all ſorts of Peſtilential or Malignant Fevers, uſed either as a prophylactick, or as a Curative... It is to be taken in all that the Patient drinks, whether at Meals, or otherwiſe, ſo many Drops at a time as to give the Ale, Beer, or Wine, a pleaſing acidity.*

The Cataplasm. *It is made of the Roots and Leaves, beaten in a Mortar, and brought to a Form of Pultife. Being applied, it is faid to cure Peftilential Carbuncles and Botches, and other like Apostems, Bubo's and Tumors...*" [Botanologia]

1722 – Curiously, despite Salmon's list of masterwort remedies above, Quincy's text from roughly the same time period, suggests that the plant was not fully utilized. Perhaps English physicians preferred to put their faith in home-grown angelica root rather than imported masterwort root (regarded as the best, and subject to adulteration because of the higher cost), and so masterwort became a secondary list herb: "*Thefe are not much unlike the* Angelica *Roots, in Flavour or Virtue. They are agreeably penetrating, and are allow'd by all to be good* Alexipharmicks...*Thefe are not fo often met with in* extemporaneouf Prefcription, *as they deferve. In the* College Plague-Water *they are an excellent Ingredient; and if they are difpens'd in it in due quantity, and frefh, give a very agreeable predominant* Tafte *to the* Compofition." [Pharmacopœia Officinalis]

1727 – Shaw's *Dispensatory of the Royal College of Physicians in Edinburgh* has the following recipe for an 'Aqua Epidemica, *Plague Water*' some of the key ingredients partly mirroring those in Thomas Cocke's '*Cordiall Antidote*' of 1665: "*Take of the roots of Mafter-wort, and Butter-bur, each four ounces; wild Valerian,* Virginia *Snake-root, and Zedoary, each an ounce and half; the leaves of Baulm, Rue, and Scordium, each three ounces; the feeds of Angelica, and thofe of Lovage, of Juniper-berries and Bay-berries, each two ounces: Bruife, flice, or cut the ingredients, as they fhall require, pour thereon two gallons and a half of French Brandy; and when they have ftood, to digeft for four days, draw off the like quantity, viz. two gallons and a half.* **

* *This compofition is not, like the Plague-water of other Difpenfatories, clogg'd with ufelefs ingredients, that afford nothing proper to the intention, by diftillation; fuch as Celandine, Carduus, Centory, Gentian, &c. but is directed with difcernment, and defign'd as a high carminative cordial in malignant cafes, or great depreffions.*" [Dispensatory, Edinburgh]

1731 – The author of the commercial work *Compleat Body of Distilling*, aimed at the wholesale market, gives a recipe for producing twenty gallons of plague-water that uses masterwort root among other ingredients. The economics of production are given in the book, which uses 23 gallons of proof spirit costing £1 18s 4d, and 1s each for the vegetative ingredients and fuel plus labour. With total costs around £2 10s the estimated sale price was £8. A handsome profit.

The vegetative ingredients were 24 handfuls each of Rue, Rosemary, Balm, Carduus, Scordium, Mint, Marigolds, Dragons, and Goats-Rue, then 3 pounds each of the roots of Masterwort, Angelica, Butterbur, and Peony, plus a pound and a half of Scorzonera root. Under the directions for making the plague-water there is this: "*It is called Plague-water, becaufe of its being a fovereign antidote or remedy* againſt it, as againſt the Cholick, Gripes, Faintings, Ill-digeſtion, &c. and has a peculiar virtue to diſpoſe one to ſleep.

The beſt ſeaſon to make it is in the month of June, when all the Herbs are at their firſt and full growth; becauſe the ſecond crop hath not the virtue and efficacy of the firſt: and it is meliorated by keeping: ſo that you muſt make as much then, as will ſerve for a whole year's ſale: or if it be kept longer, 'twill be ſo much the better; as will all goods that are made high proof: as, Plague-water muſt be in a peculiar manner, becauſe ſo great a quantity of Herbs and Roots does extremely lower, or reduce the body, or ſtrength, of your Spirits. Draw it off from your Still very gently, and no longer than proof. Be ſure you make it, and all fine goods high proof; which cauſes them to be, and taſte, far more clean, and agreeable to the palate." [Compleat Body of Distilling]

1747 – James' *Pharmacopœia Universalis*, which was aimed largely at medical professionals has yet another variation on an *Aqua Epidemia* which he attributes to the Edinburgh pharmacopoeia, although in his following comment James does not appear too keen on the remedy: "*Take of the Roots of Maſterwort and Butter Burr, each four Ounces;* Virginia *Snake Root and Zedoary each two Ounces; the Seeds of Angelica and Bay-berries, each three Ounces; the Leaves of* Scordium, *ſix Ounces; bruiſe and cut the ingredients, and pour thereon two Gallons of French Brandy; and when they have ſtood to macerate, for four Days, draw two Gallons.*
This is ſaid to be intended as a highly carminative Cordial, in very low and languid Caſes, and to raiſe the Spirits in the Plague and malignant Fevers. But in theſe Caſes I ſhould ſuſpect it of doing great Miſchiefs, for the very ſame Reaſon that it gives a temporary Relief." [Pharmacopœia Universalis]

1751 – From around the same period as James' *Pharmacopœia* we have John Hill extolling the virtues of masterwort in his *History of the Materia Medica* (again aimed at medical professionals) even to the point where he questions, like Quincy [1722] above, why masterwort is not used more often: "*The Root diſtill'd yields a very conſiderable Quantity of a fragrant eſſential Oil. It is a Sudorific, Cordial, and Carminative.* Hoffman *recommends it in a very remarkable manner in Flatulencies and Cholicks, and ſcarce leſs ardently for the promoting the Menſes, and as a Provocative. It is found to affiſt Digeſtion, and to do good in many Diſorders of the Stomach, and is given in malignant Fevers, and in the Plague itſelf. With all theſe Virtues, many of them not attributed to it by fancyful Writers, but really poſſeſſed by it, it is a Reproach to us that we do not bring it more into Uſe.*" [History of Materia Medica]

1790 – Meyrick's domestic *New Family Herbal* also abounds in masterwort's glowing properties against the plague, as well as other medical conditions: "*The root is of a cordial ſudorific nature, and it ſtands high in the opinion of many as a remedy of great efficacy in malignant and peſtilential fevers. It is likewiſe ſerviceable in diſorders of the head, ſtomach, and bowels. It is moſt efficacious when newly taken out of the ground, and the beſt manner of giving it is in a light infuſion.*" [New Family Herbal]

MEADOWSWEET – *Filipendula ulmaria*

Perennial meadowsweet, with its tufts of small, creamy white, sweet scented, flowers is a very common plant of damp meadows, river banks, and similar moist habitats, particularly those with clay soils and also moist sandy loams. Often meadowsweet is found in association with various kinds of rushes and sedges in riverine habitats, and reaches about two to four feet in height. It flowers from around May-June but can occasionally be found in bloom as late as September.

Meadowsweet went under a variety of other common local names down the centuries, some of which are highly confusing since they are also used to describe other, unrelated, plant species: Bittersweet, Bridewort, Courtship-and-matrimony, Goat's Beard, Honey-sweet, Maid-of-the-Meadow, Maid-sweet, Meadow-soot, My Lady's Belt, Queen-of-the-Meadow, and Sweet Hay. Suggestions as to the origin of the name 'meadowsweet' range from it being a corruption of 'mead sweet' – a reference to the fragrant flowers being formerly used for flavouring mead in Scandinavia – to the fact that the plant grows in moist meadows. Medicinally, the plant is better known as a source of salicin which was isolated from the plant by German and Swiss researchers in 1838, and subsequently synthesized into acetylsalicylic acid, aspirin, in the late 19th century.

1614 – The short forty eight page plague treatise by physician Jean Vigier, *Traicte de Peste*, features several remedies that contain meadowsweet – *reine de pres* in French. Indeed, the proportion of plague remedies with meadowsweet is higher in the Vigier work than virtually any other old plague text; nine references by this author's count. The following two recipes below are examples; the first is for a 'water' that Vigier describes as *'tres-finguliere'* and then one for a potion:

"Prenez racine de zedoaria, tormentille, enule campane & d'angelique de chafcune deux drachmes, canelle demy once, des trois fantaux, & d'efcorce de citron feche de chacun vne drachme & demie, feuille d'efcabieufe, pimpenelle, vinete ronde, fcordium & d'vlmaria ou roine de prés de chacune deux poignées, femence de citron, de chardon benift & de coriandre de chafcun deux drachmes & demie, qu'on infufe le tout auec trois liures de vin blanc, & d'autant d'eau de fontaine fur les cendres chaudes par l'efpace de vingt heures, qu'on diftille le tout par alembic de verre dans le bain marie, de cefte eau en prendra tous les matins vne cuillerée, ou bien on vfera de l'eau Theriacalle, ou

Prenez eaux cordiales & de la roine des prez de chafcune vne once, dans lefquelles diffoudrez Theriaque vieille trois grains, confection alkermes & de hyacinthe de chacune vn grain & demy, du bol armenie quatre grains, fyrop de limons vne once, foit fait potus, lequel on prendra au matin." [Traicte de Peste]

1665 – In a pamphlet issued by the Royal College of Physicians of London during the Great Plague – *Certain Necessary Directions* – a section on plague remedies for the Summer has the following recipe containing meadowsweet, although it is unclear if the recipe is a proper 'sweat' or given during a sweat: *"Take of the roots of Butterburre, the inner Bark of Aſh, of each a pound; Rue, Scordium, Angelica, Meadow-Sweet, Dragons, Carduus, of each three handfuls, White-Wine and Vineger of each two quarts, let them infuſe for a day or two, and after be diſtilled; adding to the reſt (if to be had) ſix handfuls of the green Rinds of Walnuts: Let the Water be ſweetned with Syrupe of Wood-Sorrel, adding to two quarts half a dram of Camphire, and three drams of Spirit of Sulphur. This Water may be given from two ounces to four."*

Although some of the plant ingredients of the above recipe do suggest that it would have been employed as a sweat, the directions for the following remedy suggest that this one would have been used as a general anti-plague preservative: *"Take of the roots of* Petaſitis, *or Butterburre ſix ounces, roots of Elecampane, Maſterwort, and Angelica, of each an ounce and a halfe, leaves of Meadow-ſweet, Scordium, Bawm* [sic]*, of each two handfuls, Rue and Wormwood of each one handful, Citron (or Limon) peel, Nutmeg, of each half an ounce, of Juniper-berries ripe and pulpey two ounces, of Carduus ſeed one ounce; All duly prepared by cutting and bruiſing, are to be mixed and put into a bag, to infuſe in ſix gallons of Ale or Beer, whereof may be drunk a draught every morning and evening; and at meals it may be mingled with ordinary Beer."* [Certain Necessary Directions]

1666 – Apothecary-physician William Boghurst provides one of the best medical accounts of the 1665-6 London plague and describes four or five compound medicines of his own including this plague water containing *ulmariae* which he tells us was used in cordials, juleps, or as an admixture: *"Rx. rad enulæ, bardanæ, petasitidis, angelicæ ana lb s.s. baccar Juniperi lb i s.s. rad. tormentil., iridis, gentianæ, acetosæ, ana ℥ ii fol scabiosæ, dracontii, angelicæ, abrotoni, scordii, cardui; frondus ulmariæ, thymi, absinthii, pimpinellæ, melissæ, galegæ; flor. tunicæ, calendulæ ana m. ii. rutæ viridis m. vi aurantior. Civil. No. xii, sem. thlaspios, acetosæ, citri, pœoniæ, ana ℥ i s.s. Cinnamoni, croci, ana ℥ i Caryophyllorum, Zedoariæ, rad. serpentariæ, ana ℥ s.s. contusa macerentur in alâ novâ et vino albo ana q.s. vel lb. x et alembico distillentur."* [Loimographia]

1666 – The 1691 publication of Willis' plague remedies in *Plain & Easie Method* gives a number of sudorific powder formulations to be given in the following liquid vehicle: *"Give any of theſe in a Spoonful or two of any Liquor, or in a Spoonful of* Sack, *with as much Peſtilential* Vinegar; *half an hour after, drink a draught of Poſſet-Drink with Medeſweet, or Woodſorrel boyl'd in it."* [Plain & Easie Method]

Curiously, neither William Salmon's *Houshold Dictionary* or *Botanologia* [1710] have any mention of meadowsweet for plague, given that Salmon wrote extensively on herbal matters relating to England. Nor is there a specific plague remedy featuring meadowsweet that immediately leaps from the pages of the voluminous *Pharmacopœia Bateana* that Salmon re-published in 1713; although the key recipes listed there would have originated with the physician George Bate (1608-1699).

1721 – In a publication that followed the 1720 Marseilles plague outbreak – *Avis de Precaution Contre la Maladie Contagieuse* – meadowsweet is referenced for its alexiterial use: "Reine des près, *en latin* ulmaria. *Cette plante paſſe pour alexitére, on ne ſe ſert gueres que de ſon eau diſtillée.*" [Avis de Precaution]

1790 – William Meyrick in his domestic herbal quotes a John Hill reference regarding meadowsweet, although specific plague-related remedies are not mentioned: "*An infuſion of the freſh gathered tops of this plant promotes ſweating, and has a ſmall degree of aſtringency. It is an excellent medicine in fevers attended with purging, and may be given to the quantity of a moderate baſon full, once in two or three hours. It is likewiſe a good wound herb, whether taken inwardly, or externally applied. The flowers infuſed in any kind of liquors impart a pleaſant taſte thereto, and mixed with mead, give it the flavour of the Greek wines.*" [New Family Herbal]

1802 – Given that Meyrick makes no personal recommendation for meadow-sweet in his domestic herbal, and that the plant was completely out of fashion as a herbal plague-cure, at least in the English herbal armoury, it is interesting to see that in the French work *Le Medecin Herboriste* from around the same period as Meyrick, that *La Reine-des-prés*, meadowsweet, still remains current in anti-plague plantlore: "Décoction contre la peste, les fièvres malignes et les maladies vénériennes. *Prenez racines de pétasite, deux onces; feuilles de reine-des-prés, de chardon-bénit, de germandrée, de chacune deux poignées; faites cuire le tout, pendant un quart-d'heure, dans, trois livres d'eau de fontaine: prenez la décoction pour boisson ordinaire dans la peste, les fièvres malignes et les maladies vénériennes.*" [Le Medecin Herboriste]

MEADOWSWEET

PENNYROYAL

PIMPERNELL

GREATER PLANTAIN

ROSE sp.

GARDEN RUE

MEADOW SAFFRON

DEVIL'S-BIT SCABIOUS

PENNYROYAL – *Mentha pulegium*

Pennyroyal is a member of the mint family that grows in damp habitats such as moist commons, around ponds, and places where water has stood all winter. It is generally a small prostrate species though sometimes occurs in a short, erect, state. In the past it was scientifically known as *Pulegium vulgare* and *P. latifolium* and, according to plantlore, derives part of its name from the Latin word *pulices*, fleas, which it was believed it could kill. Perennial pennyroyal was commonly grown in cottage gardens as a remedy for colds, when taken as a tea. It was also used as an abortifactant and recommended for nausea, gripe, expelling stones, as a diuretic, for jaundice and dropsy. The juice of the plant, it was claimed, could clear sight and was good for whooping cough in children.

Comments by Thornton [1814], however, suggest that it was not generally regarded by medical professionals as a legitimate herbal ingredient by the start of the 19th century: "*This is seldom ordered by the faculty, but is used as a popular remedy with much confidence in obstructions of the courses, or when these are attended with pain or hysteria.*" Yet Miller [1722] in the early 18th century says: "*... what we use in the Shops is generally cultivated in Gardens, where it grows large...*" which perhaps suggests that pennyroyal was more widely used in 'fringe' medicine than official practice. However, pennyroyal just occasionally appears in plague remedies, but much less often than other members of the mint family.

Like many wild plants pennyroyal had numerous local common names, among them: Brotherwort, Churchwort, Pudding Grass, Hill-wort, Lillie-riall, Lurkey Dish, Flea Mint, Organ, Organy, Pudding Herb, Pulicall. The name Pudding Herb is attributed to its use as a flavouring for black-pudding, and Pudding Grass because it was used in hogs' puddings, while Coles says: "*Penniroyall chopped and put into a bag-pudding giveth it a savoury relish*".

1583 – Royet gives the following short recipe for a plague electuary for the poor, which was to be taken at breakfast time: "*Pren pouillot auec ſuccre roſat, & en ſais vn electuaire, duquel vſeras vn peu deuant deſiuner la groſſeur d'vne chaſtaigne.*" [Excellent Traicte de La Peste]

1636 – Bradwell's *Physick for the Sicknesse, Commonly called the Plague*, written about a decade after London's 1625 plague outbreak, advocated pennyroyal among the anti-pestilential strewing herbs. Bradwell does not include pennyroyal in any anti-plague remedy within the text although he does advocate it as a good kitchen herb to be incorporated into the diet during pestilential times: "*Strew the Windowes and ledges with Rew, Wormwood, Lavender, Marjoram, Penyriall, Coſtmary, and ſuch like in cold weather; but in hot with Primroſes, Violets, Roſe-leaves, Borrage, and ſuch cooling ſcents.*" [Physick for the Sickness]

1666 – London apothecary Boghurst used pennyroyal in the following potion to help with some of the secondary symptoms in a plague attack, namely wind, belching and hiccoughs. The formulation contains several pleasantly aromatic ingredients in addition to mint-like pennyroyal: "*Rx. fol anethi, thymi, pulegii, satureii, and M i. flor fumitoriæ dul., carui, anisi, anethi, charyophyllor., nucis moschat. ana 3 i coq. in possetatu.*" [Loimographia]

1737 – The following recipe is not for pennyroyal but for garden mint, and it has been included here as an example of the occasional anti-pestilential recipe which turns up from time to time in 17th and 18th century domestic cookery and household books. This first edition of this work was published the previous year, that would have put it within twenty years living memory of the 1720 Marseilles plague outbreak which was a national concern at the time. What is interesting about the passage below is the tone of the language; this is domestic-speak, simple and instructive:

"*A Prefervative againft the* Peftilence.
Take of Rue, Sage, Mint, Rofemary, Wormwood and Lavender, of each 1 Handful; infufe them in a Gallon of the beft White-wine Vinegar; put all into a Stone Bottle clofely cover'd and pafted; fet the Bottle, thus clofed, upon warm Afhes for eight Days together. After which ftrain it through a Flannel, and put the Liquor into Bottles, and to every Quart put an Ounce of Camphire; then cork the Bottles very clofe, and it will keep fome Years. With this Preparation wafh your Mouth, rub your Temples and your Loins every Day; fnuff a little up your Noftrils when you go into the Air, and carry about you a Sponge dipt in the fame, when you defire to refrefh the Smell upon any Occafion, efpecially when near to any Place or Perfon that is infected."
[The Complete Family Piece]

1852 – By the time of Thompson's *London Dispensatory* it would seem that pennyroyal has largely been discounted as having any worthwhile herbal properties in relation to plague: "*Medical properties and uses.—Pennyroyal was formerly regarded as emmenagogue, expectorant, and diaphoretic; and was in repute for promoting the uterine evacuation, and relieving hysteria, hooping-cough, and asthma; but it is now justly considered of no value, and seldom used in regular practice.*" [London Dispensatory]

PIMPERNEL – *Anagallis arvensis*

Pimpernel is a perfect example of why it is important to read between the lines of old botanical and herbal texts, as this plant species could be confused with the *pimpinel* and *pimpernel* referred to in old cookery books. That particular ingredient was, in fact, the cucumber-like tasting salad burnet (*Poterium sanguisorba*), and not Scarlet Pimpernel as we know *Anagallis arvensis* today.

Scarlet Pimpernel grows in open ground, particularly in loam or clayey soils, and is often found around the bare margins of arable fields since this small, prostrate, annual species cannot compete with the grasses of meadows and pasture. It is sometimes found in coastal areas too, among the shingle and sandy dune systems.

The plant had many names including: Bird's Eye, Bird's-tongue, Shepherd's Clock, John-go-to-bed-at-noon, Merecrop, Orange Lily Pernel, Poor Man's Weather-glass, Shepherd's Calendar, Shepherd's Delight, Shepherd's Glass, Shepherd's Sundial, Shepherd's Warning, Shepherd's Watch, Sunflower, Tom Pimpernowl, Waywort, Wincopipe, Wink-a-peep, Weather-glass. Among this group of common names the weather and time-related instances refer to the plant's habit of closing its' flowers at the onset of damp weather and opening them when it was fine, and similarly with the noon-time attribute as the flowers need bright, or sunny, light conditions to open.

In plantlore scarlet pimpernell was supposed to be an antidote to witchcraft while on the herbal side it was believed to have the power to draw out arrows, thorns and splinters embedded in flesh. Bruised leaves laid on wounds made by rabid dogs were believed to cure problems in that department, while the juice was seen as efficacious in eye complaints. As an ingredient of Cordial waters pimpernel was regarded as alexipharmick, and the plant was also used in maniacal cases and for delerious fevers. Green [1823] writes: *"It is of a cordial sudorific nature, and a strong infusion of it is an excellent medicine in feverish complaints, which it relieves by promoting a gentle perspiration. The same simple preparation is much used among country people in the first stages of consumption."*

1665 – Lovell quotes Lonicer (1528-1586) in the entry under pimpernel saying: *"It helpeth againſt the plague, the* diſtilled water *help the bitings of mad dogs…"* while generally commenting that the plant: *"… is attenuant, and diuretic; ſaid to be alexipharmac and vulnerary; and is commended for the ſcurvy, epilepſy, melancholy, madneſs, rabies, plague conſumption, internal ulcers, the dropſy and gout, &c."* [Panbotanologia, or Compleat Herbal]

1710 – The entry for pimpernel in William Salmon's *Botanologia* is quite extensive, outlining the uses in plague as follows: "Pimpernel *is a peculiar remedy againft the Plague, and all Malignant and Peftilential Fevers, and other Contagious Difeafes...*

The Effence. *It has all the Virtues of the* Liquid Juice, *befides which it is of great force againft the Plague or Peftilence, and all forts of Malign and Peftilential Fevers, being a few times ufed, and Sweating well upon it... Dofe 2, 3 or 4 Ounces Morning and Night.*

The Decoction in Wine. *Given from 4 to 8 Ounces, it is a good remedy againft the Plague and other peftilential Fevers, and Contagious Difeafes, fo as after the taking thereof, as hot as the Patient can well drink it, they ly* [sic] *in their Beds, and Sweat for two Hours after, whereby the Poifon of the Difeafe will be expelled, this being thus ufed for 2 or 3 times.*

The Spiritous Tincture. *It has all the Virtues of the* Effence *and* Decoction, *and may be given Morning and Night from 2 Drams to 4, in a Glafs of Wine, or other fit Vehicle, againft Poifon, Plague, and Peftilence, bitings of Mad Dogs, &c...*

The Acid Tincture. *It is a potent thing againft the Plague, and all malign and peftilential Fevers, deftroying the malignity, and extinguifhing almoft in a moment the preternatural Heat... Dofe fo many drops as will make the Vehicle pleafantly fharp, and to be often repeated in the Day time."* [Botanologia]

1713 – In Salmon's reworking of the *Pharmacopœia Bateana*, that is aimed squarely at medical professionals, he has his own take on the use of one of Bate's original plague remedies containing pimpernel:

Bate.] Rx *Celandine, Rofemary, Rue, Sage, Mugwort, Wormwood, Pimpernel, Dragons, Scabiouf, Agrimony, Baum, Scordium, leffer Centory, Carduus Ben. Betony, Rof. Solif, ana* Mij. *Roots of Angelica, Tormentil, Gentian, Zedoary, Liquorice, ana* ℥j. *macerate in white Wine* ℔ *viij. for two Days; then diftil according to Art.*

Salmon.] This Water is profitable againft the Plague or Peftilence, and all Manner of malign and peftilential Fevers, &c. ... It may either be given alone from ℥ij. ad ℥iv. or uf'd as a Vehicle to convey other Medicines for the fame Intention, in letting the Sick be in Bed, Sweating well upon it". [Pharmacopœia Bateana]

1747 – From the mid-18th century we have James' *Pharmacopoeia Universalis*, aimed more at medical professionals, and where use of pimpernel in plague takes a secondary position to other recommendations: "*For a Decoction of it drank is not only commended againft the Plague, the Bites of a Viper, and mad Dog, but has alfo been found a Specific in Madnefs, after the previous Exhibition of an Emetic... If it is beneficial againft the Plague, and the bites of venomous animals, it muft be on account of its refolvent and abftergent Qualities."* [Pharmacopoeia Universalis]

1751 – Describing the scarlet flowers of the true pimpernel – *Anagallis arvensis* – John Hill's *History of the Materia Medica* describes the plant and its' qualities along with some general comments that would suggest pimpernel was more of a home remedy than used professionally: "*This Plant has the Credit of being a very powerful Sudorific, as alʃo a Cephalic and Vulnerary. It is greatly recommended by many in maniacal Caʃes, and in violent Deliriums attending Fevers: As alʃo in all hyʃteric and hypocondriac* [sic] *Complaints, and in Epileptics and Convulʃions of all Kinds. With all theʃe Recommendations however, it is ʃcarce known in the Shops, it is an Ingredient however in many of the family Medicines ʃo frequent in* England, *and it ʃeems of the number of thoʃe Plants that very much deʃerve to be try'd, in order to determine whether Fancy, or real Obʃervation and Experience have led Men to ʃpeak well of them.*" [History of the Materia Medica]

1790 – William Meyrick tells his domestic audience that: "*A decoction of it in wine, drank in bed, cauʃes ʃweating, and is a preʃervative in peʃtilential and contagious diʃeaʃes; a water diʃtilled from it is excellent for ʃore eyes. The expreʃʃed juice is ʃerviceable in the beginning of dropʃies, and in obʃtructions of the liver, ʃpleen, and reins. It brings away ʃtony and gravelly concretions from the bladder and urinary paʃʃages, and is good in conʃumptive caʃes, ulcerated lungs, and other diʃorders of the breaʃt.*" [New Family Herbal]

1822 – John Waller, a naval surgeon writing mainly for an educated domestic readership, appears to feel that perhaps more could be made of pimpernel including the use of it for treating colds: "*This plant is unquestionably possessed of valuable properties, though neglected in our practice. It has been esteemed cordial, cephalic, sudorific, vulnerary. Etmuller, with many other writers of equal note, extol it as a remedy for madness. Quercetanus was celebrated for the cure of this disease, and his treatment consisted in the use of the decoction of pimpernel, after having freely evacuated the patient with antimonial vomits and strong purges... Tragus pronounces it a remedy for the plague; directing the patient to take a moderate draught of a decoction of it in wine, then to cover themselves up well in the bed-clothes and encourage sweat. Indeed in all febrile complaints, from a common cold to the plague itself, this practice will be found most efficacious.*" [New British Domestic Herbal]

GREATER PLANTAIN – *Plantago major*

This is not from the banana plantain (*Musa* sp.) family but a rather unassuming and, to perfectly honest, visually uninspiring perennial plant species found growing along footpaths, in pastures, and on waste ground. This short plant has a rosette habit, the broad oval leaves usually smooth when young but becoming downy as they mature, and having distinctive ribs on their underside.

Along with its cousin, Ribwort Plantain (*Plantago lanceolata*), the astringent qualities of both were used in remedies to allay haemorrhages, vomiting blood, nose bleeds, bleeding wounds and similar. Of greater plantain James [1747] tells us that: "*The Root, Leaves, and Seed are used, which are heating, and drying, hepatic and vulnerary, and are principally used in all Sorts of Fluxes*," and that, "*This Plant externally used is good for Inflammations, being applied to the Parts affected. It is a Plant of excellent Use in a Diarrhæa, Hæmorrahges, and Diseases of the Eyes. The bruised Leaves are good to cleanse, and consolidate old Wounds and Ulcers. Their Juice is very proper in Intermitting Fevers, and in a Phthisis.*" Although plantain never really appears as one of the key anti-pestilential remedy ingredients some of these herbal attributes, such as the astringent quality, would have been viewed as helpful in controlling the disease in a plague victim.

1483 – A book in the incunabula collection of the British Library entitled *Regimen Contra Pestilentiam* is one of the earliest English language texts that this writer has found which incorporates plantain into plague-remedies. Notice the reference to the '*inward ſpyryte*', essentially part of the whole *Humours* notion of early medicine: "*The comfortes of the herte be theſe / Saffrō Canifey Planteyn with other herbes / they opyn the inward ſpyryte & theſe be gode among the comyn people where lightly it happeth that one is infecte of another....*
... Alſo ther is another medycyne. Take ſengrene hyl wort otherwiſe called Wyld tyme Maudelyn graſſe plantyn & a lytil rye floure and breke al theſe to gyder tyl ye ſee water come out therof Medill that water with womans mylke & gyue yt to the pacyent faſting before ſlepe & it wille werke to better for to remeue the ſwellyng: Alſo for the ſwelling whan it apped Take filberd nottes ſygges & Rewe / bruſe them to gydre & laye it upon the ſwellyng." [Regimen Contra Pestilentiam]

1569 – Almost a hundred years later the physician-surgeon John Vandernote cribbs the same lines: "*The Comfortes of the hert are saffron, camphir and plantain with other herbs.*" [Gouernance and Preservation of them that Feare the Plage]

1575 – The 1649 English version of Paré's medical and surgical works suggests a variety of medications for tackling rough and dry mouth symptoms during a plague attack. The seeds of *psillium* in the extract that follows, belong to another member of the *Plantago* family more commonly found in the southern parts of Europe: "*For the drineſs and roughneſs of the mouth, it is verie good to have a cooling, moiftening & lenifying lotion of the mucilaginous water of the infufion of the feeds of Quinces, pfilium, id eft, Flea-wurt, adding thereto a little Camphir, with the Water of Plantain and Rofes; then cleanf and wipe out the filth, and then moiften the mouth, by holding therein a little oil of fweat* [sic] *Almonds mixed with a little fyrup of Violets."* [Workes of Ambrose Parey]

1625 – In Thayre's *Excellent Treatise of the Plague* the author provides a gargle remedy with plaintain; ostensibly for dealing with those secondary symptoms of a plague attack, a painfully sore and dry throat and mouth. However, now that we know that there is a clinically different pharangeal form of plague it is to be wondered if the following remedy was, in fact, being used to treat that form of the disease:

"A Gargarifme to heale the mouth, throa [sic] *and tongue in this fickneffe, if it be fore through the heate of the ftomacke.*
R. Barley excoricated or common barley a handfull, Plantain leaues; ftrawberry leaues, violet leaues, finckfoile leaues, of either of thefe a handfull, bryer tops, halfe a handfull, Woodbine leaues and columbine leaues halfe a handfull, fhrewd and bruife thefe hearbs a little, and then boyle them in a quart of faire water which being well boyled, ftraine it forth, and put thereto Diomoren two ounces, Mel Rofarum, or Honey of Rofes two ounces: mixe thefe, and let the patient vse it often to wafh and gargarize his mouth." [Excellent Treatise]

1696 – From the same period in France there is the following remedy where herbalists and physicians turned to plantain for treating the buboes in plague rather than for internal symptoms such as the mouth and throat example above, while the astringent properties of plantain were used to treat blood haemorrhages when they appeared:

"Remedes contre les bubons peftilentiels...
         Prenez du nitre, deux gros,
De l'eau de plantain, un verre.
         Mettez fondre vôtre nitre dans l'eau de plantain, trempez de groffes compreffes dedans & les appliquez au front & aux tempes, pour appaifer la grande chaleur, si'l eft neceffaire, & confultez le Médecin là-deffus." [La Medecine Aisée]

RHUBARB – *Rheum* sp.

Rhubarb should require no descriptive outline, while the story of its physical introduction to the British Isles in the early 17th century, and the history of further varieties brought in over the following decades, would consume pages of text. Suffice it so say that once herbalists got their hands on rhubarb the plant was put through its paces as a medicinal ingredient, very much as with tobacco when samples of that species first appeared in Britain.

While rhubarb does not appear to feature in many specific plague-treating remedies it does appear in plague and pestilential 'preventatives', a couple of examples of which follow. In the text passage immediately below, from Thornton [1814], one gets a sense of how rhubarb would have been seen as useful in keeping the Humours in tune to prevent plague taking hold in the individual: "*Rhubarb is one of the best and mildest Cathartics in the whole* Materia Medica; *it operates very well on the Bile, and on all the* Viscera *of the* Abdomen, *and at the same time strengthens the Nervous Fibres. On these accounts, it is proper in weak Stomachs and Itestines* [sic]. *It is given in Substance from twelve Grains to half a Dram, and, in Infusion, from half a Dram to a Dram and a half; and, in a small Dose, becomes an excellent Alterative. It purges the Bile very effectually, and has a greater Force than any other Purgative, in opening Obstructions of the Liver... The Use of Rhubarb is, however, dangerous, when the Kidneys or Bladder are suspected to be inflamed, because it heats considerably; and for this Reason it is improper in Hæmorrhages.*" That reference to '*improper in Hæmorrhages*' would make sense for an actual plague victim, since various forms of haemorrhaging often accompanied a plague attack.

The New Dispensatory [1753] provides further insight into the trade in, and adulteration of, rhubarb in the medical world, as well as some further insight into the plant's supposed herbal qualities: "*Two ſorts of rhubarb are met with in the ſhops. The firſt is imported from Turkey and Ruſſia, in roundiſh pieces, freed from the bark, with a hole through the middle of each; they are externally of a yellow colour, and on cutting appear variegated with lively reddiſh ſtreaks. The other, which is leſs eſteemed, comes immediately from the Eaſt Indies, in longiſh pieces, harder, heavier, and more compact than the foregoing. The firſt ſort, unleſs kept very dry, is apt to grow mouldy and worm eaten; the ſecond is leſs ſubject to theſe inconveniences. Some of the more induſtrious artiſts are ſaid to fill up the worm holes with certain mixtures, and to colour the outſide of the damaged pieces with powder of the finer ſorts of rhubarb, and ſometimes with cheaper materials: this is often ſo nicely done, as effectually to impoſe upon the buyer, unleſs he very carefully examines each piece. The marks of good rhubarb are, that it be firm and ſolid, but not flinty; that it be eaſily pulverable, and appear, when powdered, of a fine bright yellow colour; that upon being chewed, it impart to the ſpittle a ſaffron tinge, without proving ſlimy or mucil-*

aginous in the mouth. Its taſte is ſubacrid, bitteriſh, and ſomewhat aſtringent; the ſmell, lightly aromatic.

Rhubarb is a mild cathartic, which operates without violence or irritation, and may be exhibited with ſafety even to pregnant women and children. Beſides its purgative quality, it is celebrated for an aſtringent one, by which it ſtrengthens the tone of the ſtomach and inteſtines, and proves uſeful in diarrhœæ: and diſorders proceeding from a laxity of the fibres. Rhubarb in ſubſtance operates more powerfully as a cathartic than any of the preparations of it."

1545 – Guido's short plague treatise makes distinction between treating the poor and rich differently on a number of occasions, but the connection between rich people with a casket, or similar, in the plague composition mentioned below is very unusual. Unless, that is, rhubarb was still such an expensive and exotic ingredient having only reached Europe in the previous one or two centuries: "*Aux riches auec caſſe, catholicon: reubarbe & agaric: qui eſt tref excellent en toutes maladies veneneuſes.*" [Briefue institution pour préserver]

1566 – A further interesting snippet that shows the value of rhubarb during the 16th century is to be seen in the passing comment from another French plague text: "*... purgez auec infuſion de Rheubarbe, s'il et riche: & s'il eſt pouure, auec Electuaire de ſuc de roſes...*"

Elsewhere in the text we find that the gentle purging qualities come into their own: "*Les debiles delicatz, femmes enceintes, enfans* [sic], *& vielles gens, ſuffire purger auec vne once de Caſſe extraicte, demye, ou vne dragme de Rhubarbe...*" That mention of *casse* again makes this author wonder if the previous Guido snippet above relates to quassia rather than a casket. [Traicte de la Peste]

1583 – For 16th century French physicians rhubarb was viewed as a plague *preservative*, Royet's work commenting: "*Auſſi la Rheubarbe tenue en la bouche, & maſchée au matin la groſſeur d'vne auellane auec vn clou de giroffle eſt preſeruatiue.*" [Excellent Traicte de la Peste]

1593 – The Kellwaye *Defensitive against the Plague* tract has rhubarb as a component for '*A good Purgation for a weake body*':

"*Rx. Fol. ſennae,*   3. iij.
   Rhab. elect,   3. j.
   Sem, aniſi,   3. ß
   Schenanthi,   Э. ß
   Aqua Acetoſæ   ℥. v.
Boyle them a little, then take it from the fire, and let them ſtand infuſed together twelue houres then ſtraine it out ſtrongly, and adde thereto.
   Syr. roſ. lax.   ℥. i.
And then drinke it as the other before."

1636 – Written about a decade after the 1625 London plague outbreak Bradwell's *Physick for the Sicknesse, Commonly called the Plague* has the following recipe for some mild preventative purging pills to be used two or three times a week to keep '... *the Bodye free from the increafe of fuperfluous humors,*' which would have been seen as undermining the general health of the individual, and their disposition towards catching plague:

"*Aloës Rofatae, unc.* I.
Rhabarbari, Croci, ana drach. 3.
Myrrha, drach. 6.
Santali citrini, drach. I.
Ambari grifei, fcrup. I.
Cum fyrupi de fucco citri, q. f.
    fiat f. a. Maffa Philularum [sic].
Make *Pils* of 8. 10. or 12. graines a piece; and take 2 or 3. at a time; either at bed time, or after the firft fleep: you may take them in *Syrup of Rofes,* or *conferve of Violets*; or if you will, in the yolke of a reare [sic] egge." [Physick for the Sicknesse]

1665 – Quoting Tentzel, Robert Lovell simply comments that: "*The* tincture *helpeth the plague, and killeth wormes the Dose is 10. or 12. dr.*" [Panbotanologia, or Compleat Herbal]

1721 – A much more complicated anti-pestilential pill recipe than that provided by Bradwell appears in anonymous plague work *Some Observations Concerning the Plague* where the author quotes a recipe from Diemerbroek:

"The *Compofition* of the *Anti-peftilential* Pills is as follows.
Take the Roots of Butter-bur, Carline Thiftle, Dittany, Angelica, Elecampane, of each half an Ounce, of Gentian one Dram and Half of the beft Rhubarb, one Ounce and Half, of Zedoary one Dram: Of tbe whiteft Agarick, Half an Ounce; Take alfo the Herbs Scordium, the leffer Centory, Rue, of each Half an Ounce: Carduus Benedictus, fix Drams; and of the Flowers of Stæchas one Dram and an half, as alfo the Seeds of Citron and Oranges, of each one Dram; of all thefe make a grofs Powder, which fteep for two or three Days in two Pounds and an half, or three Pounds of White Wine; then boil it for about a Quarter of an Hour, and ftrain it very ftrongly in a Prefs, and afterwards ftrain it again through thin Paper: In the ftrained Liquor diffolve three Ounces and an half of the beft Aloes, and three Drams and an half of clear Myrrh in Drops; Let the Moifture evaporate in a China Difh, till a Mafs of Pills can be made of the Remainder. Thefe Pills (fays the Author) *we have found to be of great Ufe in Time of the Plague.*" [Observations Concerning the Plague]

ROSE – *Rosa* sp.

As with the previous section on rhubarb, roses do not need any introduction, though herbalists of the past were quite specific about the varieties or species of rose that were regarded as herbally useful. Unlike some of the sudorific heavy-weights, such as angelica and butterbur, roses do not feature as conspicuously in anti-plague remedies. They appear, rather, to have been seen as cordial and mildly purgative as the following explanation from *The New Dispensatory* [1753] suggests: "*Rose Damascena* rofa purpurea, *The damafk rofe. This elegant flower is frequent in our gardens. Its fmell is very pleafant, and almoft univerfally admired; its tafte bitterifh and fubacrid. In diftillation with water, it yields a fmall portion of a butyraceous oil, whofe flavour exactly refembles that of the rofes. This oil, and the diftilled water, are very ufeful and agreeable cordials: Hoffman ftrongly recommends them as of fingular efficacy for raifing the ftrength, chearing [sic] and recruiting the fpirits, and allaying pain; which they perform without raifing any heat in the conftitution, rather abating it when inordinate. Damafk rofes, befides their cordial aromatic virtue which refides in their volatile parts, have a mildly purgative one which remains entire in the decoction left after the diftillation: this, with a proper quantity of fugar forms an agreeable laxative fyrup, which has long kept its place in the fhops.*

Rosa rubra, rofa rubra multiplex. The red rofe has very little of the fragrance of the foregoing pale fort; and, inftead of its purgative quality, a mild gratefully aftringent one, efpecially before the flower has opened." [New Dispensatory]

1575 – The English Paré translation from 1649 recommends a number of potions to be taken after a plague patient has eaten their meal. Quite what the intended medical outcome was, this writer is unclear: "*And in the later end of the* Meal, *Quinces rofted in the embers: Marmalate* [sic] *of Quinces, and confervs of Buglofs or of Rofes, and fuch like, may be taken: or elf this powder following.*

Take of Coriander-feeds prepared, two drams: of Pearl, Rofe-leavs [sic], *fhaveings of Hartfhorn and Ivorie, of each half a dram; of Amber two fcruples; of Cinnamon one fcruple; of Unicorn's horn, and the bone of Stagg's heart, of each half a fcruple; of Sugar or Rofes, four ounces: make thereof a powder, and ufe it after meats.*

If the patient be fomwhat [sic] *weak, hee muft bee fed with Gellis made of the flefh of a Capon, and Veal fodden together in the water of Sorrel, Carduus Benedictus, with a little quan-* [sic] *of Rofe-vinegar, Cinnamon, Sugar, and other fuch like, as the prefent neceffitie fhall feem to require.*" [Workes of Ambrose Parey]

1579 – William Bullein has the following '*fyrupus de infufone Rofarum viridium,* or greene Rofes...' recipe: "*Firft take of the infufion of yong Rofes. li.v. of Suger. li. iii. mingle theym, and make a fyrupe. This is good for the thirft in burning Agues, and to affwage the heate, it doth comforte the Stomack, Harte, and Liuer, beyne troubled with heate, it preferveth the Bodye from all corruption, and from the Peftilence, it refifteth poyfon.*" [Bulwarke of Defence]

1593 – An excessively dry mouth was a symptom of a plague attack as Kellwaye indicates: "*In this contagious difeafe, there doth chance an ulceration of the mouth which is called Aphtham, it commeth by meanes of the great interior heate which the ficke is oppreffed with, in the time of his fickness, which if it bee not well looked unto in time, it will greatly endanger the body, for remedy whereof ufe this Gargaris.*

Take, Cleane Barley, one handfull.

Wilde dayfie leaues, Planten leaues, Strauberie leaues, Violet leaues, of either one handfull.

Purflan feede, one fcruple. Quinche feede, one fcruple and halfe.

Licqueris brufed, foure drammes.

Boyle all thefe in a fuficient quantitie of water untill the water be halfe confumed, then ftraine it, and take one pinte and halfe thereof and adde thereto.

Syrop of rofes by infufion, And Syrrop of dried rofes, of either foure drams.

Diamoron, two ounces.

Mixe thefe together, and Gargaris and wafh the mouth therewith often times, being warme, and it helpeth." [Defensative against the Plague]

1625 – In the following century Thomas Thayre advocates syrup of roses as the agent to moisten the ingredients of plague preservative pills as follows:

"*The Compoftion of the Pill.*

R. Good Rubarb one dramme and a halfe, Saffron two fcruples, Trochus of Agarick one dram; of chofen Myrrhe one dram; Aloes the beft two drams, firrup of Rofes folutiue as much as Will fuffice to make them in pilles.

Take a dram of thefe pills early euery morning, for fiue or fix dayes together, taking two or three houres after them a little thinne broth, and vfe a fparing diet for thefe fiue or fix daies, and let your meat bee light & eafie of digeftion: you fhall haue two or three ftooles daily or foure in fome bodies. Not withstanding you may fafely go abroad about your bufineffe, without any inconuenience at all." [Excellent Treatise]

1713 – In Salmon's re-published work of George Bate's *Pharmacopœia Bateana* he gives the following recipe for a '*Tinctura Rofarum*, Tincture of Rofes' which appears to have been used elsewhere as a plague remedy admixture:

"Bate.] *Rx Red Rofes exungulated ʒfs. Fountain-water boiling hot ℔iifs. Oil of Vitriol gut. xxx. digeft three Hours, ftrain out and, to clear add white Sugar-candy in Powder* ʒij.

It refrigerates in Fevers, and comforts the Liver, helps Concoction, ftops Fluxes, as alfo the overflowing of the Terms. Dofe ʒij.

Salmon.] *It is a pleafing Julep to allay the Heat in Fevers, and quench Thirft; and is of good Ufe to be given in the Meafles or Small-pox, if any Flux be prefent or fear'd... It is alfo of approv'd Succefs in the Plague, and all malign or peftilential Fevers, deftroying the Root of the morbifick Caufe.*" [Pharmacopœia Bateana]

RUE, GARDEN – *Ruta graveolens* / **RUE, GOAT'S** – *Galega officinalis*

Both the garden rue (*Ruta graveolens*) and goat's rue (*Galega officinalis*) appear to have been used interchangeably in plague medicines through the centuries. Often referred to as Herb Grace in herbal works, *The New Dispensatory* [1753] refers to the garden variety as '*ruta hortensis latifolia*, Broad-leaved rue' and indicates that the leaves and seeds were used: "*This is a ſmall ſhrubby plant, met with in gardens, where it flowers in June, and holds its green leaves all the winter; we frequently find in the markets a narrow leaved ſort, which is cultivated by ſome in preference to the other, on account of its leaves appearing variegated during the winter, with white ſtreaks. Rue has a ſtrong ungrateful ſmell, and a bitteriſh, penetrating taſte: the leaves, when in full vigour, are extremely acrid, inſomach as to inflame and bliſter the ſkin, if much handled. With regard to their medicinal virtues, they are powerfully ſtimulating, attenuating and detergent; and hence in cold phlegmatic habits, they quicken the circulation, diſſolve tenacious juices, open obſtructions of the excretory glands, and promote the fluid ſecretions. The writers on the materia medica in general, have entertained a very high opinion of the virtues of this plant.*"

Regarding goats rue, named as *galega vulgaris floribus coeruleis* in the same text, the author of the *Dispensatory* seems less convinced about the positive benefits of this particular species: "*This is celebrated as an alexipharmac; but its ſenſible qualities diſcover no foundation for any virtues of this kind: the taſte is merely leguminous; and in Italy (where it grows wild) it is ſaid to be frequently uſed as food.*" However, Green's domestic *Universal Herbal* [1823], written for a domestic market suggests it was still viewed as currently used: "*Goat's Rue is esteemed as a cordial, sudorific, and alexipharmic. Mr. Boyle celebrated its virtues in pestilential and malignant diseases. The leaves, gathered just as the plant is going into flower, and dried, with the addition of boiling water, make an infusion, which being drank plentifully, excites sweating, and is good in fevers.*"

1485 – In a Latin version of the Jacobi text, *Contra Pestilentiam*, under a section headed '*De remediis peſtilētie*' there is the following entry that lists rue: "*Fiat etiā cū fumigatione herbarū infra ſcriptarū, videlicet lauribacce, iuniperi, vberiorgani, et habeť in apothecis. abſinthii, yſopi, rute, arthimeſie atǫ ligni aloes, ǭ melius valere: ſed pro paruo pecio cōparari nō poteſt.*" [Contra Pestilentiam]

1578 – Of goat's rue the Dodoens-Lyte *Herbal* says that: "*It is counted of greate vertue, to be boyled in vineger, and dronken with a litle* [sic] *Treacle, to heale the plague, if it be taken within twelue houres,*" while of the garden rue and its wild varieties the text informs readers that the wild type is stronger herbally, and that: "*The leaues of Rue eaten alone with meates, or receyued with walnuttes, and dryed figges ſtamped togither, are good againſt all euil ayres and againſt the Peſtilence and all poyſon, and againſt the bitings of vipers & Serpentes.*" [New Herbal]

1579 – Bullein comments on herb grace that: "*If a little of this Rue be ſtamped, and sodden in wyne, and drunke, it is an excellēt medicine againſt Poyſon and Peſtilence.* [Bulwarke of Defence]

1596 – Around the same period the *Rich Store-House or Treasury* medical work has the following similar advice: "*A very good Medicine for the Plague... Take in the morning faſting, one dry Fig, one Wallnut, and 4 or 5 leaues of hearbe-grace, chopped all together very ſmall, and eate them, and drinke afterwards a good draught of white or claret Wine: If it be a woman with child, leaue out the hearbe-grace. This hath been proued."* [Rich Store-House or Treasury for the Diseased]

1625 – In his plague treaty Thomas Thayre has the following remedy for the 'Commons' as he terms them, although one would have thought figs were well beyond the pocket of many poor people: "*R. Figges ſeauen or eight in number, Rue one handfull, the kernels of ten or twelue Walnuts cleane picked from their skinnes, foure or ſixe ſpoonfulls of good Vinegar, beate theſe together in a Mortar, and keepe it cloſe in a boxe, and eate thereof euery morning, and it is good to defend thee from the infection.*" [Excellent Treatise of the Plague]

1629 – London apothecary John Parkinson, writing a few years after the 1625 London plague outbreak, comments in his gardening book that: "*Rue or Herbe grace is a ſtrong herbe, yet vſed inwardly againſt the plague as an Antidote with Figs and Wall-nuts, and helpeth much againſt windy bodies: outwardly it is vſed to bee layde to the wreſtes of the hands, to druie away agues...*" [Paradisus Terrestris]

1640 – A decade later Parkinson tells readers of his *Theatrum Botanicum* that: "*Both ſorts of Rue (that is) the garden and the Wilde, as* Dioſcorides *ſaith, doe heate, burne, and exulcerate the skin. It provoketh urine and womens courſes, being taken in meat or drinke. The ſeed thereof taken in wine is an Antidote or Counterpoiſon againſt all dangerous medicines, or deadly poiſons: the leaves hereof taken either by themſelves, or with Figges and Walnuts, is called* Mithridates *his counterpoiſon, or Mithridate againſt the plague, cauſeth all venemous things, as well as of Serpents, to become harmeleſſe...*"

Of goat's rue Parkinson continues: "*It is no leſſe powerfull and effectuall againſt poiſon then* [sic] *the plague or peſtilence, or any infectious or peſtilentious fevers or diſeaſes, that breake forth into ſpots or markes, as the meaſells, purples, and the ſmall pocks, in all which it is admirable, what effects it worketh, both to preſerve from the infection, and to cure them that are infected, to take every morning ſome of the juice thereof, as alſo to eate the herbe it ſelfe, every morning faſting, but it will be the more effectuall if the juice be taken with a little good Treakle and ſome Tormentill rootes in powder, mixed with* Cardus bendictus *water, or with ſome vinegar and fine Bole-armonicke, and Treakle in the ſaid water, and preſently to ſweat two houres thereupon, which it cauſeth alſo in ſome ſort it ſelfe, and may be uſed as well*

when any is infected, as when any feare themſelves to be infected with the plague..."
[Theatrum Botanicum]

1665 – The pamphlet issued by the Royal College of Physicians of London during the Great Plague of London, *Certain Necessary Directions*, has a recipe for a preservative 'Inward' medicine as the text terms them: "*To ſteep Rue, Wormwood, or Sage all night in their drink, and to drink a good draught in the morning faſting, is very wholſom, or to drink a draught of ſuch drink after the taking of any of the preſervatives will be very good.*" [Certain Necessary Directions]

1665 – Coinciding with the College's *Directions* was Lovell's *Panbotanologia*, where the author refers to other sources of information when writing about goat's rue: "Boiled *in vineger and drunk with treacle, it prevents the plague: eaten in* fallads *with oile, vineger, and pepper, it preventeth venemous infirmities, and cauſeth ſweat... The leaves of rue eaten with the kernells of walnuts, or figges ſtamped helpeth evill aires, the peſtilence, ſo* Ger. Berg. Untz. Palm. Mind. Dur." [Panbotanologia]

1666 – Medicinal herbs could become costly as supply-demand became an issue in plague times, as can be seen in the following excerpt from Boghurst where he suggests brewers should produce herbal root beers: "*... the Brewers should brew all their drink with alexipharmicall Herbs, roots etc., instead of Hops, but they must bee, such as may bee had in good quantities. The Ingredients are these, viz., Carduus, Scordium, Scabious, Aron leaves, Burdocks, Angelica, Southernwood, pimpernell, wormwood, Goats' rue, Marigold, Elicompane leaves or roots, Juniper berries, enough to bee had, Walnut leaves or the tender boughs, and many more might bee reckoned, but all these may bee had in quantityes great enough, and yet are not any of them soe ill tasted but that they are all tolerable in drink, though they were all brewed together. I have forborne to name rue, because it is soe strong in taste and smell, and it is so dear and scarce to bee had, especially in a plague. They that can abide it and have it growing may make use of it. Scordium also is commonly dear, which may be left out if not easily had, though it is one of the best of them.*" [Loimographia]

1668 – Markham's 17th century answer to Mrs. Beaton's home-help works around two centuries later was the domestic-orientated *English Housewife*, and among the various home remedies for plague is a '*Drawing poultice*' for plague sores: "*Take* Smallage, Mallowes, Wormwood, *and* Rue ſtamp them well together, and fry them in oyl Olive, till they be thick, plaiſter-wife apply it to the place where you would have it rife, and let it lie untill it break, then to heal it up, take the juice of Smallage, Wheat-flower, *and milk, and boile them to a pultis, and apply it morning and evening till it be whole.*" [English Housewife]

1693 – Dale's professional oriented *Pharmacologia*, a materia medica in Latin for the English market, draws on Schroeder's pharmacopoeia for garden rue: "*Alexipharmaca, Cephalica, ac Nervina eſt. Uſus præcip. in variis morbis, Peſte, aliiſque affectibus malignis præfervandis ac curandis...*" When it comes to information on the virtues of goat's rue Dale's source remains unidentified: "*Celeberrimum eft Alexipharmacum ac Sudorificum, Venenum imprimis peftilentiale infigniter difcutiens. Ufus ejus præcipuus in Petechiis expellendis, aliifque Morbis peftilentialibus, ipfaque Pefte curanda...*" [Pharmacologia]

1710 – Salmon *Botanologia* provides information on a number of rue-based preparations with an orientation towards treating plague and pestilence:

"The Sallet. *The Herb it felf is eaten, being boiled with Flefh, as we ufed to boil and eat Cabbage and Coleworts; it is alfo eaten as boiled* Spinage, *and other Sallets, with Pepper, Salt, Vinegar and Oil; and fo being eaten, it is faid to be excellent againft all forts of Poyfons, and the malignity or infection of the Plague or Peftilence, or the Bitings of venomous Creatures, &c.*

Liquid Juice. ... *it is alfo effectual againft Vegetable Poyfons, as alfo the malignity of the Plague, or Peftilence it felf, and the infection of other Peftilential or Contagious Difeafes... for it both preferves from infection, and perfectly cures fuch as are infected; Dofe, two or three Spoonfuls every Morning fafting in a Glafs of Generous Wine, and as much at Night going to Bed, by way of prevention; but for Cure it ought to be given in* Angelica *Water, in the fame, or larger Dofe, according to the Age and Condition of the Patient, and to be repeated as often as the exigency or vehemency of the Difeafe requires.*" Salmon also comments that the essence has same virtues as juice but '*acts more fpeedily and powerfully*'.

"The Electuary. *Take Pouder of Goat's-Rue four Ounces, Zedeory in Pouder, Contrayerva, Virginia Snake-root, all in fine Pouder, of each one Ounce, Saffron, Cochenele, of each two Drams, Rob of Goat's-Rue twenty Ounces, mix and make an Electuary.* It is good againft all forts of Poyfons, both Vegetable and Animal, as alfo againft the Plague, and has indeed all the Virtues of the *Liquid Effence* and *Juice*; Dofe, from one Dram to two Drams, in any proper Vehicle.

The Acid Tincture. *It has all the Virtues of the Juice and Effence, whether inwardly taken, or outwardly applied; it is a Specifick againft the Plague, and alfo againft all other Malign and Peftilential Difeafes, and cures the moft violent burning Fevers in a very fhort time, taking away the violence of their burning heat in the fpace of an Hour, Dofe, fo many drops as to make the Vehicle pleafantly fharp, and to be given in all that the Patient drinks.*" [Botanologia]

1713 – In Salmon's republished work of the Bates' pharmacopoeia, *Pharmacopœia Bateana*, Salmon provides his own recipe for a "*Tinctura Ruta*, Tincture of Rue' for use in plague treatment:

Rx. Rue carefully dryed ℥iv. Salt of Tartar ℥ij.
Grind them together in a hot Iron Mortar, put them into a Glafs Matrafs, to which affufe of the beft rectified spirit of Rue ℔*iij. mix and cover the Veffel immediately with a blind Head, luting well the Juncture.*

Digeſt all in a very gentle heat for fourteen days, ſhaking the Glaſs once a day, then being fine, decant off the Tincture, and keep it for Uſe.

It is uſed chiefly in the Plague and other malign Diſeaſes, to cure them, and preſerve from them, to expel Poyſon, extinguiſh Luſt, cure the Pleuriſie, the Colick, Bitings of Serpents, mad Dogs, or other venomous Creatures. [Pharmacopœia Bateana]

1714 – The de Ville reprints of Gaspar Bauhin's work comment on the garden rue: *"… elle eſt excellente contre les venins, & s'en ſert d'ordinaire pour ſe preſerver de la peſte la mangeant en ſalade…"* and of goat's rue: *"Mangée cruë en ſalade, ou cuite dans le potage, elle a une vertu admirable pour preſerver de la peſte; le même arrive à ceux qui boivent tous les jours ſon ſuc à jeun dans le vin; elle eſt auſſi bonne contre les venins, & les morſures des ſerpens priſe de même, comme auſſi dans les fiévres peſtilentielles & pour tuer les vers dans le corps."* [Histoire des Plantes de Europe]

1721 – The French language text on the 1720 Marseilles plague outbreak, *Avis de Precaution Contre las Maladie Contagieuse de Marseilles*, has the following snippet about rue's uses: *"Ruë. Fevilles, fleurs, & graine. C'eſt un grand remede contre la Peſte. On infuſe toutes ſes parties dans le vinaigre. On applique les fevilles ſur les tumeurs externes. On prend la graine pilée au poids de trente grains dans une cuillerée de vinaigre, ſon efficace eſt plus grande que celle des feüilles pour preſerver. On ſe ſert auſſi de l'eau diſtillée, & de l'extrait. La plante ſeche eſt bonne auſſi à brûler en parfum. Son huile diſtillée trois gouttes avec un peu de ſucre dans une cuillerée d'eau de chardon-benit pour preſervatif, ſept ou huit gouttes pour provoquer la ſueur, & ſervir de curatif. On ſe ſert auſſi de ſon ſel."* [Avis de Precaution]

1722 – The Quincy text, *Pharmacopœia Officinalis*, does not suggest that the following recipe for an '*Acetum rutaceum, Vinegar of Rue*' is used for plague fever, but when one considers that only a couple of years earlier news of the Marseilles plague outbreak (1720), and the fear that plague could arrive in Britain, may well have seen householders prepare the potion as a general stand-by sweat remedy that could double-up in times of a plague attack:

"Infuſe the Leaves of Rue and Scordium, *that is,* Water-Germander, *pick'd from the thick Stalks, ana m. iii. Juniper-Berries and Angelica Roots, ana ℥ ii. Zedoary and Sevil Orange-Peels, ana ℥ i. in the beſt Vinegar ℔ viii. let them digeſt a Month, and then preſs the Vinegar from the Ingredients, which keep for uſe. This is not preſcrib'd, or kept in the Shops, but is ſo eaſily made by any private Family, and is ſo good a Medicine to procure Sweat upon any threatnings of a Fever, or upon a Surfeit; that it is very well worth any one's making, and keeping by him. It may be given from half a Spoonful to two or three, in any convenient warm Liquor. And if the Patient is kept warm with Clothes, it cannot fail of raiſing a Sweat; and it is the beſt* Succedaneum *to Treacle-Water in the World, where that cannot be had for a ſudden occaſion."* [Pharmacopœia Officinalis]

1747 – *Medicina Britannica* refers to rue (it appears the garden variety) as the *Country Man's Treacle* and recirculates the wine infusion for domestic plague use, saying the plant: "... *contains much exalted Oil and volatile Salt. It cuts, thins, and difcuffes Humours, refifts Poifons, peftilential and contagious Difeafes... Its Infufion in Wine, drank, is a good Prefervation againft the Plague, malignant and epidemic Difeafes.*" [Medicina Britannica]

1747 – The comments on rue in James' *Pharmacopœia Universalis* seem to suggest that it was a garden cultivar of the wild plant that was used: "*Goats Rue. It grows by River Sides, and marfhy Places, in feveral Parts of Italy but with us only in Gardens; it flowers in June and July. It is a moft celebrated Alexipharmic and Sudorific, and moft powerfully difcuffes peftilential Poifon; its principal Ufe is in expelling petechial Eruptions, and other peftilential Difeafes, and in curing the Plague itfelf; it is good in the Meafles, and Epilepfies of Children, and the Stings of Serpents, and deftroys Worms, even by external Application.*" [Pharmacopœia Universalis]

1751 – Writing for a more medically professional readership in his *History of the Materia Medica*, John Hill says of goat's rue: "*Galega has the Reputation of being a very great Alexipharmic and Sudorific, good againft the Plague, malignant Fevers, and the Bites of venomous Animals. We have the largeft Accounts of its Virtues from the Italians, who have it native among them, and who eat it raw and boiled, and make a kind of Tea of it, as well as ufe it in their Decoctions and Ptifans given to People in Fevers of all Kinds. With us it is not at prefent in any Efteem; it is only kept in the Shops as an Ingredient in fome Compofitions.*" [History of the Materia Medica]

1761 – In *Primitive Physick* preacher John Wesley's tips in times of plague included medical cuisine too: "*Or, a little of the tops of Rue with Bread and Butter, every Morning.*"

1790 – Meyrick's *New Family Herbal* tells his domestic audience that of the garden rue (*Ruta graveolens*) variety: "*The tops of the young fhoots contain the greateft virtues of any part of the plant. An infufion of them may be taken in the manner of tea, or they may be beaten into a conferve with three times their weight of fugar, and taken in that form. The infufion is good in feverifh, complaints, it raifes the fpirits, promotes perfpiration, and expels the matter which occafions the difeafe.*" For goat's rue (*Galega officinalis*) he says that: "*The leaves of this plant gathered juft as it is going into flower, and dried, with the addition of boiling water, make an infufion, which being drank plentifully excites fweating and is good in fevers.*" [New Family Herbal]

SAFFRON – *Colchium autumnale / Crocus sativus*

These days Saffron is probably more readily associated with cookery and the kitchen rather than medicine, being an ingredient of many Indian dishes and, of course, saffron bread. There are about two dozen, or so, *Colchium* species found in a range from India to North Africa and in Europe, and the bulbs or corms of many them are highly toxic. Then there is the *Crocus sativus*, or garden crocus, which is generally the source of the fragrant commercial culinary spice these days.

The whispy stigmas of saffron flowers were the parts of the plant used and frequently made their way into remedies for hysteric depression, alleviating pains and spasms, and obstructions of the body, while Roger Bacon believed that saffron could prevent the effects of old age when mixed with certain medicines. In immoderate quantities, however, the stigmas can kill, as can the bulb if consumed. However, tiny amounts of the bulb were employed in treating conditions like dropsy, and used along similar lines to squill (*Scilla* sp.).

In England it was mainly the Meadow Saffron or Autumn Crocus that was preferred and used in herbal medicine, while continental physicians more than likely employed *Crocus sativus* which grew extensively in warmer parts of Europe, southern Spain in particular. Because the stigmas are highly time-consuming and therefore expensive to harvest it was not unknown for unscrupulous merchants to substitute the bright orange-yellow petals of marigold (*Calendula officinalis*) for saffron, and if the product was sold in powdered form it would have been very difficult to identify the adulterated item.

Preferring limestone-based and somewhat dry soils our own meadow saffron is a perennial plant of meadows and woodland, where it grows from about six to twelve inches in height, and where its long purple-to-white flowers make a colourful display from around August to September when they are in bloom.

1575 – Ambroise Paré, in the English 1649 version of his earlier work, has saffron added to the diet of the plague victim in a form of medicinal-cuisine; presumably to prevent obstructions within the patient's digestive system: "*If the patient at anie time bee fed with ſodden meats, let the brothes bee made with Lettuce, Purſlain, Succorie, Borage, Sorrel, Hops, Bugloſs, Creſſes, Burnet, Marigolds, Chervil, the cooling Seeds, French-Barlie and Oat-meal, with a little Saffron, for Saffron doth engender manie ſpirits, and refiſteth poiſon. To theſe opening roots may bee added to avoid obſtruction; yet much broth muſt bee refuſed by reaſon of moiſture.*" [Workes of Ambrose Parey]

1578 – Bullein, in his *Dialogue* plague text has the following recipe for some pills against the plague: "*The beste Pilles generallie vnder heauen, and is thus made. Take the beste Yellowe Aloes, twoo vnces, Myrrhe and Saffron, of eche one vnce,*

beate them together in a Morter a good while, putte in a little sweete wine, then rolle it vp, and of this make fiue Pilles, or seuen of one dragme; whereof take eurie daie next your harte a Scruple or more, it will expulse the Pestilence that daie, &c."
[Dialogue Against the Fever Pestilence]

In his other work of the same period, *Bulwarke of Defence* [1579], Bullein identifies the mixture above as being for 'pills of Ruffi', which were a commonly made anti-plague remedy, and further comments: "*Thefe Pilles be much ufed of the Phifitions if they be taken in fommer, and if you take them in the plague time, then you muft put unto them as much Bolearmoniake as of Aloes.*"

Elsewhere in *Bulwarke* there is this 'defensive' potion with saffron: "*The pouder of Safron. x. graynes, Walnuts, twenty graynes, Figges. 3. ij. & fixe Sage leaues ftamped together, with 3. j. of Pimpernel water, and three graynes of Mithridatum: kepe this in a clofe glasse, and eate thereof in the morning twelue graynes, and this will defende the receyuer thereof, from the Peftilence.*" [Bulwarke of Defence]

1617 – Regarding anti-pestilential Ruffi pills Woodall, in his East-India Company ship's surgeon's manual describes the activities of three key ingredients: "*... and fitting for the peftilence and plague, doe rather preuent infection, then cure the infected: for by reafon of the aloes the body is freed from excrements, by myrrha from putritude, and by Saffron the vitall faculties are quickened...*" [Surgions Mate]

1625 – In a medicine for the poor, Thomas Thayre has this saffron-containing defensive potion captioned '*For the Commons*', the ingredient list also including those anti-pestilential herb stalwarts angelica and tormentil:

"*Another good preferuative that defendeth all from Infection.*
Take kernels of Walnuts three ounces, Rue one ounce and a halfe, fine bole Armoniacke one ounce, roote of Angelica and Turmentill of either an ounce, good figs three ounces, Myrrhe three drame, Saffron foure penny-worth.
Let thefe be beaten a good fpace in a Mortar, then put thereto two or three fpoonfulls of good Vinegar, and as much Rofe-water, and incorporate them well together, eate hereof as much as a hazell nut in the morning, and at any other time of the day going where the infection is, and bee free from all infection.*" [Excellent Treatise on the Plague]

1629 – From Parkinson's general domestic gardening work, *Paradisus Terrestris*, the author unashamedly promotes English meadow saffron as the valued herb: "*The true Saffron (for the others are of no vfe) which wee call Englifh Saffron, is of very great vfe both for inward and outward difeafes, and is very cordiall, vfed to expell any hurtfull or, venemous vapours from the heart, both in the fmall Pockes, Meafels, Plague, Iaundife, and many other difeaffes, as alfo to ftrengthen and comfort and cold or weake members.*" [Paradisus Terrestris]

1710 – Salmon's *Botanologia* provides his readers with the three following saffron-containing plague herbal recipes:

"The Infufion in Wine. *...Take choife Canary a Quart; of the beft Englifh Saffron an Ounce; mix, and Infufe in a cold Digeftion for 40 days of more, fhaking the Bottle once every Day; let it fettle, and decant the clear: to the Fæces put a Pint more of the fame Wine; digeft 3 Weeks, fhaking the Bottle once every day; and then letting it ftand till it is fettled, decant the clear into the former Infufion, and fo keep it for ufe.* Sixty or eighty Drops may be given of it at a time in a Glafs of Wine... againft the Measles, Small Pox, Spotted Fever, and the Plague or Peftilence... both as a Prophylactick or Prefervative, and as a Curative...

The Tincture. *It is made by mixing an Ounce of the beft English Saffron with a Quart of Spirit of Wine in a cold Digeftion for 40 days, fhaking the bottle once every Day; if half an Ounce of Cochinele be added to the Saffron, and 2 Ounces of the beft rectified Spirit of Salt, digefting cold, as aforefaid, and then decanting the clear Tincture; it will be of double ftrength, and have all the Virtues of the Pouder and Infufion in a fuperabundant manner.* It chears the Heart, refifts Melancholy, and enervates the Venom of Malignant Difeafes: being profitable againft the very Plague or Peftilence it felf... It may be given from half a Dram to a Dram, or 2 Drams, according as the exigency requires, in any proper Vehicle.

Take Angelica Water, Treacle Water, of each an Ounce; Syrup of Clovegilliflowers 6 Drams; Ticture [sic] *of Saffron a Dram and half; mix for a Dofe,* to be given in Mailignant Fevers, Small Pox, Meafles, Plagues, &c." [Botanologia]

1710 – Given that Salmon's *The Family Dictionary: or Houfhold Companion* is aimed at a domestic readership it seems amazing that he would include the complex 'water' recipe below, while a number of the ingredients, including citrus fruits, would have been expensive to procure: "Bezoardick Water... *Take Methridate fix pounds;* Virginia *Snake-Root, Contrayervya, Zedoary, of each fix ounces; Cloves, Mace, Nutmegs, Cubebs, Cardamoms, Caraways, Bayberries, Juniperberries, Gentian, Winters, Cinnamon,* Jamaica *Pepper, black Pepper, Ginger, of each three ounces; Saffron, Cocheneel, Limon-peels, Orange peels, (the yellow only) of each two ounces; Rofemary and Lavender-flowers, Angelica, Bawm* [sic]*, Mint, Peniroyal, Sage, Savory, Thyme, Sweet Marjoram, of each three handful; Spirit of Wine three Gallons, bruife what are to be bruifed and digeft all together for fourteen days; then put thereto white Wine four Gallons; diftil all in an Alembick with a Refrigeratory, and draw off three Gallons of pure Spirit, which referve; then change the Receiver, and draw off two Gallons more, which make into a Syrup with treble refin'd Sugar, to which add the firft diftilled Spirit, fhake them well together, and let them ftand till they are fine.*

It is good againft Poifon, Plague, fpotted Fever, Small Pox, Meafles, and all forts of Malignant Fevers... It is a very great Cordial, good againft Sadnefs and Dejection of Mind, revives all the Spirits and makes merry a fad and drooping Heart. Dofe, two fpoonfuls or more, now and then upon any Illnefs or other occafion."

Salmon also gives domestic households a recipe for a specific tincture of saffron which was also '... *good againſt the Infection of the Plague*' among other medical conditions: "*Digeſt in two Quarts of our* Aqua Bezoartica, *two ounces of Saffron for the ſpace of ſix days; then the Tincture being ſtrained out, keep it cloſe ſtopped for uſe. You may take of this two drams at a time in a Glaſs of Wine, or other convenient Liquor.*

It is wonderfully Efficacious in Chearing and Comforting the Heart, Concocting the Crude Humours of the Breſt, helps the Jaundice, is good againſt the Infection of the Plague; and is of ſingular Validity in driving out the Small Pox..." [The Family Dictionary]

1713 – For professional medics Salmon had the following saffron-containing *Aqua Cordialis*, cordial water, in his *Pharmacopœia Bateana*: "Rx *The outward Rind of Limons freſh.* ℥iv. *of Oranges* ℥iij. *Cinnamon* ℥ij. *Nutmegs, Mace, Calamus Aromaticuſ, Coriander, ana* ℥j. *Cubebs* 3vj. *Cloves, Seeds of Carduus ben. Cardamums unhusk'd, ana* ℥ ß. *Saffron* 3j. *Spirit of Wine* ℔vj. *Fountain-water* ℔xij. *infuſe and diſtil according to Art.*

It is good againſt the Plague, Peſtilence, and all Manner of malign Fevers; in which laſt Caſes it ought, to be given in ſome cooling Vehicle, with three or four Drops of Spirit of Sulphur in it." [Pharmacopœia Bateana]

1722 – While the Salmon remedies above appear to use saffron unabashed, Quincy's *Pharmacopœia Officinalis*, from about the same period, does highlight the expense of saffron in the following extract on the plant. There is no particular mention of its relevance in plague treatments although one could perhaps regard the phrase '... *any Species of Fevers'* as including plague fever. It is somewhat at variance with Salmon, considering that the 1720 Marseilles outbreak had only recently occurred. Canary, incidentally, was a sweet wine from the Canary Islands that was popular at the time: "*It is certainly one of the greateſt Cordials of any Simple the whole* Materia Medica *ſupplies; and as effectually promotes a* Diaphoreſis, *which makes it hardly ever omitted in extemporaneous Preſcriptions, for any Species of Fevers.*

The Dearneſs of this Commodity makes ſome draw out its Tincture *for a Syrup, or to uſe by it ſelf; and afterwards dry and powder it for uſe. That which has not been ſo ſerved is almoſt of a red Colour when powder'd; and upon but juſt touching it with any Moiſture, will ſtain extremely* yellow. *The Colour which it gives in Tincture, tho deep and fine at firſt, will fade with keeping, and the ſooner, as the* Menſtruum *is more acid; for this reaſon, that which is order'd with* Treacle-Water, *and ſeems beſt fitted to anſwer the Intention of an* Alexipharmick, *is hardly ever made: it generally is done with* Canary, *and ſuch Wines, which are moſt remote from Acidity...*" [Pharmacopœia Officinalis]

SCABIOUS, FIELD - *Knautia arvensis* / DEVIL'S-BIT - *Succisa pratensis*

Scabious appears in quite a number of plague-related herbal remedies and both the field scabious and devil's-bit scabious appear to have been used, possibly interchangeably. Perennial field scabious (*Knautia arvensis*) is the bigger of the two plants, and prefers a slightly dry, sandy, habitat, regularly finding a home in the dry banks of open fields and hilly areas, although it will be found growing in abundance in arable fields too. The plant can grow up to about two feet tall, is often found growing in clumps, and has conspicuous, bright lilac to blue, stalked flowerheads that bloom from about July to September-October.

Herbally the plant is astringent, and was used for fevers, coughs, asthma, and epilepsy among other ailments. The plant was supposed to be useful in treating scabby eruptions and skin irritations, hence the common English name – according to the Dodoens-Lyte herbal. However, field scabious had a whole variety of localized common names that ranged from Bachelor's Buttons or Blue Buttons and Bluecaps, to Gipsy Flower, Lady's Cushion, and Pincushion.

Devilsbit, sometimes Devils-bit, Scabious, is a much smaller and more slender plant than the field scabious, and is often to be found growing in fields and meadows, lanes and along roadsides, and tolerates lower altitudes and valleys as well as hillside habitats and alpine rocks. It prefers clay-like soils and sandy loams, and grows to about one and half feet tall. Again, this scabious species has conspicuous pale violet to white coloured flowers, but they are smaller than those of the *Knautia*. Occasionally the plant is referred to as *Scabiosa succisa* in old botanical works.

As with *Knautia*, tradition has it that this bitter-tasting plant was used to treat scabby conditions and scabies, and also went under a variety of other common names such as Blue Bonnets, Blue-heads, Blue-kiss, Gentleman's Buttons, Hardhead, Woolly and Stinking Nancy. Incidentally, the plant yields a yellow and green dye and, by some accounts, was used in tanning, which would probably account for the bitter taste mentioned.

1527 – Brunschweig's *Vertuous Book of Distillation* (originally published in 1500) recommends a scabious 'water' and that: "*Every morning drink of this water an ounce fasting, is good for the pestilence.*"

1551 – Under a section describing preservative medicines against the plague Benoit Textor's *Maniere de Preserver de la Pestilence* says:

"*Aucuns ſimples medicaments fort louez en cecy, & bien faciles à trouuer.*
      Suc de Colondula ceſt ſolſie beu à la quãtité de deux ou trois onces, le malade eſtãt bien couuuert. Ou de verbenaca, ou de ſcabieuſe, par lequel la malice du venin eſt dechaſſee dedens douze heures ſelon aucuns. Racine de tormentille prinſe en poudre

le poix dune drachme ou de deux, auec quelque liqueur idoine, ou bien le fuc dicelle
tiré auec vinaigre, ou la decoction. La cichoree tant le ius que la decoction dicelle en
efté. Bolarmeni amené au premier traicté. Ie pourrois faire vn plus grand denombremēt
de telz remedes fimples que dautres nont pas omis, mais ie me contente de ceux cy,
comme des meilleurs." [Maniere de Preserver]

1578 – Calling devil's bit scabious *morsus diaboli* the Dodoens-Lyte herbal has
an inward decoction of the root, but also the pounded plant used as a poultice
for plague sores: "*The decoctiō of Deuels* [sic] *bit, with his roote, boyled in wine*
& drōken, is good againſt al difeaſes, that Scabious ſerueth for, & alſo againſt the
Peſtilence... Diuels [sic] *bitte freſſh* [sic] *and greene gathered, with his roote and*
floures pounde or ſtamped, and layde to Carboncles, Peſtilential ſores and Botches,
doth ripe and heale the ſame." [New Herbal]

1578 – In his anti-plague treatise *Dialogue Against the Fever* Bullein suggests
the following potion with a scabious element to its construction for occasions
when the reader has not taken their regular anti-pestilential pills:

"*Potion for mornings when pills not taken:*
None better than this: take Theriaca, *of the making of* Andromachus, *ij Scruples,*
whiche is a Triacle incomperable [sic]*, paſſyng againſte bothe poison and Pestilence;*
and the Antidotari *of* Mithridatis *1 Scruple; bole Armoniacke, prepared, half a Scruple;*
and of the water of distilled Roses, Scabious and Buglosse, of eche one vnce, mingled
together." [Dialogue Against the Fever]

1579 – The content of Bullein's other contemporaneous work of the time,
Bulwarke of Defence, takes the form of a dialogue between the fictional characters
Marcellus and Hilarius who discuss various aspects of medicine and remedies
of all sorts, in an attempt to make the information more accessible to readers.
Marcellus comments that "*Our fyelds doe grow full of Scabious, euery mere and balke*
is full of it in June: Some dooe uſe this Herbe for the Peſtilence, commonly called
the plague. Wherefore doe the learned men, as Dioſcorides *and* Galen *ſay it is good.*"

The second character, Hilarius, adds: "*Scabioſa ſo named of old tyme, becauſe*
it is giuen in drynke inwardly, or oyntmentes outwardly, to heale Scabbes, ſores,
corruption in the ſtomacke, yea is moſt fryend among all other herbes in ȳ tyme of the
Peſtilence, to drink the water with Mithridatum a mornynges. It ſtoppeth the bloudy
flixe, and Hemeroides, and an oyntment made thereof, healeth Antrax, *called the*
Carbuncle, or hotte burnyng Peſtilent botch, or fearful ſore." [Bulwarke of Defence]

1585 – From Italy we find Castore Durante's herbal advocating scabious root
for plague in the following reference: "*L'Acqua ſtillata alla fin di Maggio dalle foglie,*
& dalla radice della Scabbioſa, tagliate minute uale à le poſteme, & à la ſtrettezza
del petro beuuta al peſo di tre oncie la mattina à mezo dì, & la ſera, & è gran rimedio
alla peſte, a i ueleni alla toſſe, & alla rogna." [Herbari Nuovo]

1593 – Kellwaye has an interesting variation on some of the walnut-fig-rue 'preventative' remedies seen elsewhere here, in that one ingredient is washed in scabious water prior to mixing:

"Kernels of walenuts and figs, of either four ounces. Leaues of rue, one ounce and halfe. Tormentill rootes, iiij. drammes. Rinde of fowre Citrons, one dramme. Right Bolarmoniake, vj. drammes. Fine Mirre, ij. fcruples. Saffron, one fcruple. Salte, halfe a dramme. Syrop of Citrons and Lymons, iiij. ounces.
The herbes, rootes, and rindes muft be dried, the nuttes muft be blanched, and the bolearmonyake muft be made in fine powder, and then wafht in the water of fcabios, and dried againe, you muft pounde the figges and walenuts in a ftone morter feuerally by them felues very fmall, all the reft muft be made in fine pouder, and fo mixe them all together in the morter, and then adde therto Syrop little and little, and fo incorporate them all together: you may giue this in the fame quantytie, and in like forte as the other before [ED – Given in a quantity the size of three beans and taken every two or three days]."

A further formulation in the Kellwaye text is for *'A Iulep to quench thirft and refift vennenofitse'*. The recipe is multiple-choice like many of the others in his text, and includes those key herbal thirst quenchers sorrel and orange:

"Take. Water of Scabios, Borrage, Sorrell, of either two ounces. Sirrop of Limons, Sowre Citrons, and the Iuice of Sorrell, of either one ounce.
Mixe all thefe together, and giue thereof, as caufe requireth."

In another part of the book Kellwaye deals with treating plague sores and the need for a 'maturative' to help the botch suppurate. Commonly, roasted onions are mentioned in plague books but among the remedies Kellwaye advocates is one that contains scabious, although the root or foliage is not specified:

"Take. White lillie rootes, Enulacompone roots, Scabios, & Onyons, of either two ounces. Roaft all thefe together in a cole leafe, or a wet paper, then pound them with fome fweete butter and a little Vennes triacle, whereunto doe you adde fome Galbanum, and Ammoniacum diffolued in Vinegar, and ftraine from the feffes and dregges, and fo mixe them altogether and applie it, renuing it twife a day." [Defensative against the Plague]

1640 – John Parkinson, the London apothecary-botanist, outlines scabious for the readers of his *Theatrum Botanicum* then seems to suggest that the juice was being used in a plague 'sweat': *"Now that we are come to handle the varieties of the Scabioufes, they are fo many, that I know not well how to marfhall them into any good method or order, yet that I may endeavour it the beft I can, I thinke it fitteft to ranke them into foure Orders; that is, firft, of fuch forts as grow in the Meddowes and Pafture grounds; next of thofe that grow in the Woods, and upon hills and high grounds...."*

Foure ounces of the clarified juice of Scabious *taken in the morning fasting, with a dram of Mithridatum or Venice Treakle, doth free the heart from any infection of the plague, or pestilence, so as upon the taking thereof they sweate two houres in their beds at the least, yet after the first time taking, let them that are infected take the same proportion againe, and againe if need be, for feare of further danger, the greene herbe also bruised, and applied to any Carbuncle or Plague fore, is found certaine by good experience, to dissolve or breake it within the space of three houres..."* [Theatrum Botanicum]

1665 – The pamphlet put out by London's Royal College of Physicians during the Great Plague gave the following remedy for breaking a bubo: *"Take of Scabious two handfuls, stamp it in a stone morter, then put into it of old Swines greafe falted two ounces, and the yolk of an egge; stamp them well together and lay part of this warm to the fore."* [Certain Necessary Directions]

1665 – Contemporaneous with the College pamphlet was Lovell's *Panbotanologia* where he appears to draw on knowledge from Minderatus: *"It helpeth the bitings of ferpents, and ftinging of venemous beafts, drunk and applied the juice drunk efpecially with treacle caufeth fweat, and confumeth plague fores, being given often in the beginning. So Mind. It is thought alfo to help all peftilent feavers."* [Panbotanologia]

1668 – Markham's family household self-reliance work, *English Housewife*, provides its readers with a quick and easy potion that could be rustled up in a country kitchen and used in the time of a plague attack: *"Take* Featherfew, Malef-lot, Scabious, *and* Mugwort, *of each a like, bruife them and mix them with old ale, and let the fick drink thereof fix fpoonfulls, and it will expell the corruption."* [English Housewife]

1693 – The Latin language *Pharmacologia* from Samuel Dale draws on Schroeder's German pharmacopoeia for *Scabiosa Officinalis*: *"Alexipharmaca ac Pulmonica eft. Ufus præcip. in Pleurifi, Angina,Tuffi, Pefte, fiftuloifis ulceribus. Extrinfecus in Scabie, Pruritu, impetigine, &c."* For devils-bit, *Morfus Diaboli, Succifa Officinalis* the comment is: *"Alexipharmaca & Vulneraria eft, ut Scabiofa, cum qua & reliquis facultatibus convenit, &c."* [Pharmacologia]

1710 – Devil's-Bit, says Salmon's *Botanologia*: *"... is a peculiar Antidote againft the Plague, Poyfon and Bitings of Venomous Beafts,"* and the liquid juice of the plant: *"Taken inwardly to two, three or four ounces, according to the Age of the Patient, it is powerful againft the Meafles, Small-Pox, Calenture, malign and fpotted fever, as alfo againft the Plague, and all malign and peftilential Difeases..."* while Salmon recommends the acid tincture as being: *"... a fingular Medicament againft the Plague and Spotted Fever: it abates the Heat, and quenches the Thirft..."* [Botanologia]

1721 – Published a year after the deadly 1720 Marseilles plague outbreak *Avis de Precaution* lists scabious among *'Plantes béfoardiques'* to be used in the event of plague with instructions that they were *'Pour en ufer en manière de thè'*. For many centuries bezoar – a concretion from the intestines of land-animals – was supposed to have magical healing powers and regarded as possessing alexipharmic properties. Other substances with similar properties were called bezoars too, hence the term 'plant bezoars' used in this French text. Here it looks as if both the field and devils-bit forms of scabious were being used, if we take *morfus diaboli* to be the latter species: "*Mèlez enfemble du fouci, du fcordium, de la fcabieufe, de la meliffe, du marrube blanc, du morfus diaboli, de là véronique, feüilles & fleurs fechées à l'ombre, & bien épluchées de chacune une poignèe, de la menthe de jardin, du dictam de crète, des fleurs de fureau, de camomille, & de bétoine de chacune demi poignèe: Tout coupè menu, & gardé dans un fac de papier bien bouché.*" [Avis de Precaution]

1722 – Although Quincy's *Pharmacopœia Officinalis* identifies scabious as pectoral, and for use in asthmas and pleurisies his text seems to suggest that scabious was hardly in use in England, let alone in time of plague. Perhaps the fondness for tisanes among French herbalists could explain why both scabious species appear in the previous 1721 work: "*This is mighty common in the Fields, and flowers in* July. *It has a great Character amongft many* Difpenfatory *Writers, but feems to grow much out of ufe, to what it has been... It has a Place alfo amongft the* Alexipharmicks... *There is a compound Syrup under its Name in the Shops, but of very little ufe.*" [Pharmacopœia Officinalis]

1747 – A couple of decades on and James' *Pharmacopœia Universalis* does give an indication of scabious being used by English physicians in plague, as well as for many other medical complaints it should be said. James refers to a specific scabies species *Scabiosa pratensis hirsuta quae Officinarum*, the entry for the plant being: "*The Leaves are ufed. It is alexipharmic and pectoral; and is principally ufed in Apoftems, for the Pleurify, Quinfey, Coughs, Afthma, the Plague, and fiftulous Ulcers. It is externally ufed for cutaneous Eruptions, as the Itch, and Leprofy. Scabious is bitter, and gives a faint Tincture of red to the blue Paper, which gives us Reafon to believe, that it contains a Salt refembling Sal Ammoniac, and joined with a great Quantity of fetid Oil, and Earth; for, by the Chymical Analyfis, befide feveral acid Liquors, a great deal of Sulphur and Earth, and a little urinous Spirit, and volatile concrete Salt, are obtained from it. Scabious is good to promote Expectoration, when the* Bronchia *and Veficles of the Lungs, are ftuffed with a glutinous, and condenfed Phlegm. This is a good Remedy in malignant Fevers, Small Pox, and Meafles, after the Ufe of Antimonial Medicines.* Tabernaemontanus *fays, that the Juice of Scabious mix'd with a little Borax and Camphire, takes away the white Spots, that are often feen upon the* Cornea *of the Eye.*" [Pharmacopœia Universalis]

SCORDIUM – *Teucrium scordium*

The Scordium that is present in a large number of plague remedies is the plant that today we refer to as water germander. The native habitats are wet meadows, ditches and marshes, and although distributed across Europe water germander is rather more rare in Britain.

Appearing visually like a mint, with square stalks and small purplish to rose coloured flowers that appear from around June to September (and reminiscent of dead-nettle blooms), this herbaceous perennial has a creeping, running, root-stock. Depending on the habitat it may only grow a few inches tall, but in optimum conditions may reach from one to two feet in height.

The fresh leaves are bitter and have a garlic-like odour which sometimes saw the plant referred to a *Garlick Germander*. According to plantlore the plant was used by countryfolk in a pulverized form to expel worms, while fomentations of the whole plant were used in gangrenes, and like many species in herbal plantlore, water germander was supposed to tackle nips and bites of venomous beasts and serpents. While many old references extol the virtues of scordium in plague the following, if ambivalent, entry in the *New Dispensatory* [1753] probably best sums up the plant and its' waning popularity over time: *"This is a ſmall, ſomewhat hairy plant, growing wild in ſome parts of England, though not very common; the ſhops are generally ſupplied from gardens. It has a bitter taſte, and a ſtrong diſagreeable ſmell. Scordium is of no great eſteem in the preſent practice, notwith-ſtanding the deobſtruent, diuretic, and ſudorific virtues which it was formerly celebrated for. It enters ſix officinal compoſitions, and gives name to three of them, though not the moſt valuable of their ingredients."*

1570 – Very curiously, Estienne's *Maison Rustique* suggests that water germander is one of the medicinal plant species that country house owners and farmers in France should consider growing in a medicinal herb garden, while offering the following note on its' potential use: *"Il a ſemblable vertu que la germandree d'eu, contre la poiſon & la peſte: & outre cela ſa decoction priſe en breuuge par certains iours quariſt les fieures tierces..."* [Maison Rustique]

1585 – Castore Durante's herbal extols scordium in plague: *"Vale lo ſcordio mirabilmente nelle febri peſtilentiali, & conuienſi ancora alla preſeruatione & curatione della peſte."* [Herbari Nuovo]

1640 – For apothecary-botanist John Parkinson the appeal of scordium in the middle of the following century still remains strong: *"It is a ſpeciall ingredient both in Mithridate and Treakle, as a counterpoiſon againſt all poiſons, and infections either of the plague or peſtilentiall or other Epidemicall diſeaſes, as the ſmall pockes, meaſels* [sic], *faint ſpots, or purples: and the Electuary made thereof, named* Diaſcordium, *is effectall for all the faid purpofes: and befides is often given, and with good fucceffe before the fits of agues, to divert or hinder the acceffe, and thereby to drive them away."* [Theatrum Botanicum]

1665 – Lovell's *Panbotanologia* lists scordium under a generalized *Germander* or *Chamedrys* banner and also includes one or two other germander species which may have been similarly used, though it is unclear whether all the properties are being ascribed to water germander: *"Hereof is made* diafcordium, *ferving... againft the pocks, meafles, purples or any peftilent ficknefs, before it hath univerfally poffeft the whole body...* Tinctura fcordii, *or the* Tincture of fcordium, *hath the virtue of the herb; but is better for cold ftomacks, and old bodies. Alfo it is good in malignant, peftilentiall and venemous difeafes...* Compound water of fcordium, *preferves from ill aire; a tafter being taken in the morning. Alfo it's good in peftilentiall, venemous and malignant difeafes.*

Diascordium, *or the* Electuary of fcordium...
It's more temperate than London treacle, mithridate, or venice treacle, and is therefore more generally ufed in all ages, fexes and difeafes, that require such a remedy. [ED – venemous, pestilential and malignant diseases among others]. *It's of generall ufe in all feavers efpecially when fleep is wanting. the D. in feavers is* fcr. *1.* drach. fem. *or* drach. *1. if need.... To perfons infected with the plague, it may be given thus.* R. unc. *2. in the juyce of forrel, or wood forrel,* unc. *1. of the juyce of citrons.* drach. *1. of diafcordium,* fcr. *1. of the cordiall fpecies of pretious ftons* [sic], *vineger,* unc. *1. M. give it to drink; and repeat it, as occafion requires."* [Panbotanologia]

1722 – The Quincy *Pharmacopœia* lays more emphasis on general sweating properties for scordium: *"It is juftly efteem'd a good* Alexipharmick, *and makes an excellent Ingredient in all Compofitions of that Intention; as well as in that grand Medicine in the Shops, which has its Name from it,* Diafcordium; *tho in Prefcription it moft frequently bears the Name of its Inventor, and is call'd* Confectio Fracaftorii. *There is alfo an excellent* compound Water, *which derives its Name from this Simple, and is titled in the Shops* Aqua Scordii Compofita. *It is good in all kinds of Fevers, and helps both to raife the Spirits as a Cordial, and promote Sweat, or the Eruption of fuch Humours, as do often critically terminate fuch Diftempers."* [Pharmacopœia Officinalis]

1731 – The *Compleat Body of Distilling* is an early 18th century commercial book on making alcoholic beverages mostly for sale. However, it does contain the recipe below for a *London Plague-water* the virtues of which the author says: *"I could not here omit giving you the prefcription of this moft excellent compofition; for certainly nothing can be more regularly and judicioufly compos'd, or more effectually levell'd again the peftilential Miafmata, than this great Alexipharmic is."*

The recipe contains many of the plants that you should be familiar with by now, and also scordium, water germander. What is interesting about this book is that it takes the plague-water out of its normal medicinal status and treats it as a commercial product: *"Take green Walnuts, one pound and a half, Angelica-root half a pound, Angelica-leaves, Rue, Sage, Scordium, each three handfuls, Nuts, long Pepper, Ginger, Camphire, Gentian, each an ounce and a half, Snake-root, Conthryerva* [sic], *Elecampane, Zedoary, Vipers-flesh, each four ounces; Venice Treacle and Mithridate, each four ounces, white wine Vinegar two pounds, proof spirits two gallons; macerate, and distil, s. a."* [Compleat Body of Distilling]

1747 – James appears to suggest that scordium is still in current use mid-18th century: *"Water Germander. It grows in marshy Places, flowering in June. The Herb is used. It is alexipharmic, and sudorific, and is principally used in the Plague, and pestilential Disorders. It is recommended for malignant Fevers, for Obstructions of the Liver and Spleen, for purulent and mucilaginous Infarctions of the Lungs, and for destroying Worms."* [Pharmacopœia Universalis]

1747 – For domestic households *Medicina Britannica* informs readers that: *"Many running and pestilential* Buboes *and malignant Ulcers have been cured by Use of its Powder only..."* [Medicina Britanncia]

1751 – In Hill's *History of the Materia Medica*, aimed at medical professionals and students, the suggestion is that although scordium had contributed towards plague remedies in the past it has virtually fallen into disuse or is extant as a medicinal herb, and is in varianace with James above: *"Scordium has long had the Credit of being a very great Sudorific and Alexipharmic. It is prescribed by all the medical Authors who have treated of it in malignant and pestilential Fevers, and in the Plague itself. It is an Attenuant and Dissolvent, and is on that Account recommended also in Obstructions of the Liver and Spleen, and in all Disorders of the Lungs arising from a viscous or a purulent Matter lodged there. It is said to destroy Worms. Externally it cleanses foul Ulcers, and applied by Way of Cataplasm mitigates Pain. It is at present however, not used in any of these Intentions alone, and is merely kept in the Shops as an Ingredient in the* Confectio Fracastorii, *which though it contain several Medicines of more Virtue, yet takes its Name* Diascordium *from it."* [History of the Materia Medica]

1790 – In the later domestic herbal from William Meyrick the suggestion is that scordium still remains a valuable herbal ingredient: *"The leaves when rubbed betwixt the fingers emit a strongish smell, somewhat resembling that of garlic. They are recommended as being excellent in malignant and pestilential fevers, and in weakness and laxities of the stomach and intestines."* [New Family Herbal]

SCORDIUM

COMMON SORREL

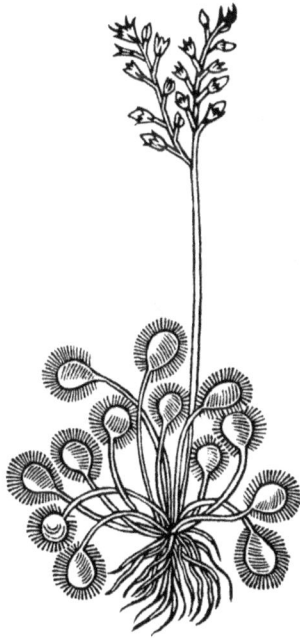

SUNDEW

SWEET CICELY

TORMENTIL

SWEET VIOLET

WALNUT

WOOD SORREL

SORREL, COMMON – *Rumex acetosa*

Readers interested in the subject of food may well be familiar with sorrel from that perspective, since the leaves have a refreshing rhubarb-like taste and flavour that makes it an excellent kitchen ingredient. Leaves for that purpose generally come from a domesticated variety of the plant (*R. scutatus*) while the wild common sorrel may also be used. In France sorrel foliage was extensively cultivated for culinary use, while a couple of old references suggest that the seeds, which have a milder acid taste, were ground up and added to bread flour or, in one other case, that the ground seeds were similarly used in their entirety for making a bread. In French texts sorrel is seen as *oseille* or *ozeille*.

Herbaceous common sorrel is a perennial plant of meadows, pastures, grassy hedge banks or open woodland. Growing generally from one to two feet in height this sorrel is a dioecious species, having the male and female flowers on different plants. However, the flower clusters are insignificant and generally most people walk past sorrel totally unaware of its presence and the tasty acidic leaves on offer.

Sorrel has a refreshing taste and so found its way into thirst-quenching remedies, but was also used to promote urinary discharge, for its refrigerant qualities, and against scurvy. Regarding the leaves and roots, *The New Dispensatory* [1753] says: "*Their medical effects are, to cool, quench thirft, and promote the urinary difcharge: A decoction of them in whey affords an ufeful and agreeable drink in febrile or inflammatory diforders; and is ftrongly recommended by Boerhaave to be ufed in the fpring as one of the moft efficacious aperients and detergents. Some kinds of fcurvies have yielded to the continued ufe of this medicine... The roots of this plant have a bitter aftringent tafte, without any acidity: They are faid to be deobftruent and diuretic; and have fometimes had a place in aperient apozems, to which they impart a reddifh colour. The feeds are fomewhat aftringent, without acidity or bitternefs: They are recommended in diarrhoeas and dyfenteries, but have long been ftrangers to the fhop.*"

According to the naval surgeon John Waller [1822] it appeared to have favour when balancing the 'Humours', that former fundamental Galenic concept behind much healing: "*Wherever a high degree of tendency to putrescence exists in the humors, the juice of sorrel will be found to possess considerable antiseptic powers, and on that account the antient physicians employed it in purid and malignant fever*".

1551 – In an section on plague diet in Benoit Textor's treatise there is the following recipe for a type of green medicinal sauce, if one can term it like that: "*De vinaigre, de perfil verd, item dun peu de mente verde, de marioleine, de ferpolet, de bafilic & de femblables: item dune bonne quantité de ius doxeille, le tout broyé & pafsé enfemble, en adiouftant de canelle ou de quelque autre efpece, ou de la poudre maintenant dite. Si on y adiouſte vn peu de ſafram, combien quil changera la coleur, il ſera bien bon."*

Elsewhere in the work there is a further '*Potion ou bruuage bien parable & facile, que eſt de grāde uertu ſelon lopiniō & experience daucuns ſauvants medecins de Paris*' with sorrel as the lead ingredient: "*Prenez doxeille lōgue ſauuage: haſchez la bien menu, & ſi voulez, broyez la vn peu. Faites la tremper leſpace de vingt & quatre heures en bon vinaigre blanc autant quil en fault pour la couurir. Faites la diſtiller à petit feu en alembic de verre ou de bonne terre. Incontinent que le perſonnage eſt frappé du mal, baillez luy en à boire vn bon verre. Apres cela ſil peult, il ſe pormenera iuſques il ſue. Lors il taſchera de dormir ou pour le moins il ſe repoſera dedans le lict bien couuert.*" [Maniere de Preserver de la Pestilence]

1563 – A medical treatise on plague originally written at the start of the 16th century appears in the larger, mid-century, *Medendis Humani Corporis* of Petri Bayri. The text makes frequent references to Avicenna, Galen to a lesser extent, and other earlier medical writers, and among the cures for plague sores, '*Cura bubonis*', is the following one which calls for sorrel water. The ingredients contain powdered camphor (in *trochisi* form) but also a syrup and water content, and it is not clear to this writer whether the final product is used as a salve, liniment-type remedy, or a fomentation or wash: "*Si ſtatìm apparuerit bubo, tunc incipe à bezoardicis, vt ſuprà dictum eſt de peſte, ſcilicet primò exhibere 3.j. trochiſci de camphora cum ſyrupo acetoſitate citri & aqua acetoſae vel roſarum rub. & ſi appareat per duos vel tres dies pòſt, tu haec iam feciſti.*" [Medendis Humani Corporis]

1575 – The great Paré had another way of using sorrel in the plague (cf. *Medendis* above) for a drink or julep for infected patients that mixed both bitter and acid elements, as the English translation [1649] shows: "*The uſe of the Julip following is alſo verie wholſom. Take of the juice of Sorrel well clarified, half a pinte; of the juice of Lettuce ſo clarified, four ounces; of the beſt hard ſugar, one pound; boil them together to a perfection; then let them be ſtrained & clarified, adding a little before the end a little vineger, and ſo let it be uſed between meals with boiled water, or with equal portions of the water of Sorrel, Lettuce, Scabious and Bugloſs.*" [Workes of Ambrose Parey]

1570 – Unsurprisingly, sorrel appears in the French *Maison Rustique*, and had been known as a culinary foodstuff in that country for centuries. However, the text also reminds readers that sorrel had use in plague: "*L'ozeille tremper au vinaigre, & mangee au matin à ieun, eſt preſeruative de peſte.*" [Maison Rustique]

1579 – Bullein tells his readers in *Bulwarke* that sorrel is: "*... known in the North as 'Sower Dockes'*" and that: "*...it ouercōmeth Choller, and is much commended. It helpeth the yellow Jaundice, if it be drunke with ſmall wine or ale, and alſo quencheth burning* *feuers: to eate of the leaues euerye Morninge in a Peftilence time, is mofte holfome, if they be eaten fafting."*

The instructions in a recipe for *Syrupus de Acetofa* are: *"Take of the iuice of Sorel clarifyed li.iii. of Sugar clarified li.ii. make of thefe your Syrupe. This Syrupe is good agaynfte Cholerike perfonnes, and Tertian Agues, the burning of the ftomache and the hart. it* [sic] *is a fynguler remedye in plages, and in Agues of corrupt humours."*
[Bulwarke of Defence]

1593 – Kellwaye has a penchant for multiple-choice remedy formulations in his *Defensative*. The following is for *'An excellent good water againft the plague, and diuers other difeafes, which is to be made in May or Iune'*, however no dose is given:

"Take Angelica, Dragons, Scabios, of either three handfuls. Take Triafandalum, fixe drammes. Ebeni, two drammes. Saffron, halfe a fcruple. Lettis feede, one dramme. Waters of rofes, Buglos, and Sorrell, of either fixe ounces. Vinegar, two ounces.
Boyle them all together a little."

In another remedy for a *'cordiall julep'*, sorrel in the form of a 'water', takes a mid-stage role as an admixture: *"Take Waters of Endiue, Purflane, and Rofes, of either two ounces. Sorrell* [sic] *water, halfe a pinte, Iuyce of Pomgarnards, and for lacke thereof Vinegar, four ounces. Camphire, three drammes. Sugar, one pounde.*
Boyle all thefe together in the forme of a Julep, and giue three or foure fponefuls thereof, at a time."

A further cordial *'... to be given where great heate is'* combines sorrel with syrup of lemons and borage:
"Rx. Conferuæ Borag, 3. iiij.
Conferuæ fol, acetofæ, ʒ. j.
Bolarm, veri, 3. j.
Manus chrifti cum perlis, ʒ. j.
Syr, de limonibus, q. 5. Miffe.
You muft often times giue of this, where great heate is."
[Defensative against the Plague]

1614 – *Les Secrets du Seigneur Alexis Piemontois* is a French translation of a household medical book published in Italian during the middle of the previous century. The contents can perhaps be best equated as a Italian-French equivalent of Gervase Markham's *English Housewife* and contains almost every conceivable household remedy – from whitening teeth to making green ink, to tips on alchemy. Within it is a section on the plague and its remedies, and a couple of the cures involve sorrel. The first of these, *'Autre compofition fort bonne en temps de pefte...,'* contains sorrel seeds and camphor, and also pearls – which possibly gave the patient a feeling of misplaced well-being as they spent heaps of money on their treatment in their time of need: *"Prens vne once de la meilleure teriaque, demie once de ius de limon, vn ſcrupule de ſafran, des deux ſortes de perles, de coral rouge, & de ſemence d'ozeille de chacun demie drachme, deux grains de camfre: meſle le tout enſemble tresbien auec deux ou trois gouttes de vin blanc odoriferant, & en fais onction, de laquelle en eſtendras quelque quantité ſus vn drap de foye cramoiſie, l'appliquant ſur le coeur tout chaud, le renouuellant au matin & au ſoir.*" [Les Secrets du Seigneur Piemontois]

1625 – Published during the key London plague outbreak prior to the 1665 epidemic Thayre's plague treatise presents the following two juleps to *'help dryness and thirst'*:

R. Water of roſes, Enduie, and Bugloſſe, of each foure ounces. ſorrell water foure ounce, good vinegar foure ounces, iuice of limons foure ounces, ſugar one pound, boyle them a little ouer a ſoft fire, which done, and cooled againe, giue to him a little thereof to drink, the quantity of two or three ſpoonefulls at a time.

Or another…
R. Sirrup of Endiue comp. ſir. of ſorrell of each three ounces. Water of Roſes, and Bugloſſe, of each one ounce, ſirrup of limons two ounces, mixe them.
Let the patient haue ſometimes, or as often as he is dry, one ſpoonefull of this ſirrupe, which is very good: and this ſhall ſuffice for the amending of his heate and drineſſe. Giue him to eate ſometimes of a limon with ſugar, or of a Pomegranate, which are both very good." [Excellent Treatise of the Plague]

1633 – A somewhat unusual combination of sorrel with oil of sulphur – which the author commends as being *'good'* – can be found in Thomas Harper's *Storehouse* and appears to be a type of sweating potion: "*Take water of Sorrell, of Roſes, of Cardus Benedictus, of each ꝫ j. Oyle of Sulfer Э j. mixe them together, and give it to the Patient as ſoone as hee feeleth himſelfe ſicke, and lay him downe to ſweat one houre, and let him not ſtirre that he maye ſweate the better, then dry him well with warme clothes, and let him reſt two houres, and then if he have liſt to eate, let him have a cullus made of a Henne…*" [Storehouse of Physical & Philosophicall Secrets]

1636 – Coghan's *Haven of Health*, which went through several editions, was first published in 1584, and has similarities in content with Sir Thomas Elyot's earlier *Castell of Health* [1534]. Without having viewed all the previous editions of both works the following snippet on sorrel could well have been circulating for almost a hundred years: "*Sorrell is cold in the third degree and drie in the ſecond; the leaves being ſodden do looſe the belly. In a time of Peſtilence, if one being faſting do chew ſome of the leaves, and ſuch downe ſome of the juice, it marvellouſly preſerveth as a new practiſer called Guainerins doth write; and I myſelfe have proved in my houſhold, faith Maſter Eliot in his Caſtell of health. Which practice proveth that greene ſawce is not onely good to procure appetit, but alſo wholſome otherwiſe againſt contagion.*" [Haven of Health]

1665 – The pamphlet issued by the Royal College of Physicians of London during the Great Plague offered this piece of advice about sorrel as a preservative against the plague: "*In all Summer plagues it ſhall be good to uſe Sorrel-sawce to be eaten in the morning with bread, and in the fall of the Leaf to uſe the juyce of Barbaries* [sic] *with bread alſo.*" [Certain Necessary Directions]

1710 – Salmon's professional-oriented *Botanologia* suggests that certain medics still contemplated the use of sorrel in plague, which is somewhat at variance with the text extract from Quincy's *Pharmacopœia* [1722] that follows afterwards. Of sorrel, Salmon advises: "The Juice or Effence. *Given from a Spoonful to three, either alone, or with ſome proper Vehicle, they cool the heat of Fevers, quench Thirſt, and take away the malignity of Infectious and Peſtilential Diſtempers...*

The Decoction in Water or Wine. *It has all the Virtues of the Juice or Effence, but nothing near ſo powerful, and therefore may be taken to half a pint or more at a time.*

The Diſtilled Water. *It is good for all the purpoſes aforeſaid, but muſt be taken in larger quantities, as four or ſix Ounces at a time, ſweetned* [sic] *with the Syrup aforegoing.* Take Juice of Sorrel ſix Ounces: of the Diſtilled Water two Ounces: Syrup of Sorrel one Ounce and a half, mix them. *Of this the Patient may take two or three Spoonfuls now and then, in any hot diſease, or burning Fever: it cools the Inflammation and heat of the blood in Agues, reſiſts peſtilential diſeaſes...*" [Botanologia]

1722 – Quincy's *Pharmacopœia* seems to offer a different perspective on sorrel at about the same time, suggesting that medical professionals were re-evaluating acid sorrel in cases of plague: "*It is well known in our common* Sallets, *and is little elſe us'd. It is acid and grateful to the* Stomach, *quenches* Thirſt, *allays the Heat of* Choler; *and, as* Schroder, *with many others, ſay, reſiſts Putrefaction, and is of great uſe in Peſtilential Fevers. But a better* Theory, *and Experience, now renounces ſuch Practice, and convinces us, that ſuch things by retarding and chilling the Juices too much, give them an opportunity of fermenting, and running into Corruption and Putrefaction. It is a grateful and ſalutary Ingredient in a Summer* Sallet, *if eat in moderation: But it is little taken notice of in medicinal Preſcription.*" [Pharmacopœia Officinalis]

SUNDEW – *Drosera rotundifolia*

The inclusion of sundew (Rosa solis in the old days) among the plague plants here is simply because it is such an unusual species and will perhaps be better known to readers as one of the carnivorous plants; capturing flies on its sticky leaves which then close up as the prey touches them. The sticky substance on the leaves contains an enzyme allowing the plant to digest its insectivorous meals.

Rosa solis is a perennial plant, growing up to about six inches tall, having spoon-shaped leaves and a preference for wet, spongy, peat habitats where there is plenty of sunshine; typically marshes, bogs and upland moors. It is often found in association with species such as Bog Pimpernel, Butterwort, Bladderwort, Bog Myrtle and Lousewort.

Not altogether common, but not rare, the number of plague remedies in which sundew features as an ingredient is not extensive either, so just regard it as a curiosity in that respect. Sundew sap is almost corrosive and was traditionally used to treat corns and warts, while the plant in milk was used to remove freckles and other skin blemishes.

In the Dodoens-Lyte herbal, which calls it *Sonne Dewe*, the following explanation for the plant's other name, Rosa Solis, is given: "*This herbe is of a very ftrange nature and maruelous: for although that the Sonne do fhine hoate, and a long time thereon, yet you fhall finde it always moyft and bedewed, and the fmall heares thereof always full of little droppes of water: and the hoater the Sonne shineth upon this herbe, fo much the moyftier it is, and the more bedewed, and for that caufe it was called Ros Solis in Latine, whiche is to fay in Englifhe, the dewe of the Sonne, or Sonnedewe.*"

1665 – Lovell, in *Panbotanologia*, references Schroeder's German pharmacopoeia regarding the plant: "*Some commend it as good againft the plague, and wounds, as alfo the epilepfie.*" [Panbotanologia]

1710 – Salmon in his domestic *Family Dictionary* offers up the following 'water' that was to be administered at the onset of a plague attack, rather than used as a 'preventive': "Water *for the* Plague. *Take of Savory, Rofemary, Bawm* [sic], *Pimpernel, Dragons, Thyme, Scabious, Agrimony, Bettony, Angelica, Pellitory of* Spain, *Carduus, Wormwood, Rosa Solis, Southern-Wood, Pellitory of the Wall, Red Sage, Setwall, Mother of Thyme, Devils-bit, and Tormentil, of each a good handful, with a few Sprigs of Rue, and Walnut-leaves, or green Walnuts; bruife all well, and let them infufe in a fufficient quantity of White Wine three Days and Nights, keeping the Pot they are in very clofe, yet often fhaking it, that the Ingredients may move in the Wine; then diftil the Wine and Herbs, and keep the Water that is drawn off clofe ftopped in Glafs Bottles, where the Sun may come to them.*

...Ten ſpoonfuls of it may be taken Blood-warm, to prevent the Plague; and this muſt be done when the firſt Symptoms of it appear. And upon taking it, walk about till you ſweat again, for the ſpace of an hour, not eating or drinking after it for the ſpace of two hours, or more; and then go to Bed upon it. If it chance to cauſe you to vomit, it is a ſign it has taken Effect." [The Family Dictionary]

1713 – In Salmon's contemporaneous professional publication *Pharmacopoeia Bateana* there is the following Ros Solis plague preservative infusion attributed to Bate: "Rx Freſh Roſa Solis, or Sun-dew, M. iv. *Nutmeg, Aniſeeds, Coriander, red Roſes dried, A. ʒſs, Galangal, Ginger, Cloves, A. 3ſs. Liquorice ʒj; Cardamoms, Zedoary, Grains of Paradice* [sic], *Calamus, Aromaticus, A. 3ſs. yellow Sanders 3j, red Sanders, Cinnamon, A. 3vj. beſt Aqua Vitae ℔ vj. infuſe for ſome Days, then ſtrain thro'* Hippocrates's *Sleeve, and add of the beſt white Sugar ʒxij.* It is good againſt a Phthiſis or Conſumption of the Lungs, a Tabes or Pining, &c. It comforts the Heart, Liver, and Ventricle; eaſes the Pain of the Head, and is a Preſervative from the Plague, &c . Doſe, 3j. - iij. &c." [Pharmacopoeia Bateana]

1731 – The author of the commercial *Compleat Body of Distilling*, which was aimed mainly at 'wholesale distillers' as the book's author puts it, describes three methods of making rosa solis water, the virtues of which he says: "*It comforts the heart, liver, and ventricle, eaſes the pain of the head, and is a preſervative from the Plague.*"

The first of the recipes here actively uses heat in the production process while the second method uses passive infusion: "*Take Roſa Solis pick'd clean one pound and a quarter; Cinnamon, Cloves, Nutmegs each an ounce; Marigold-flowers a quarter of a pound, Caraway ſeeds three ounces, proof-ſpirits three gallons, Water two gallons; draw off your proof-ſpirits from the Still, and infuſe in a quart of liquor, Liquorice ſlic'd four ounces, Raiſins ſton'd one pound, red Sanders four ounces; infuſe upon hot aſhes to a due extraction of their virtue; ſtrain, and diſſolve therein white Sugar one pound and a half, which when cold mix with the proof goods for uſe.*

Take Roſa Solis cleans'd four handfuls; Cinnamon, Nutmegs, Caraway and Coriander ſeed, each one ounce; Cloves, Mace, Ginger, each three drachm's; Cardemums, Cubebs, Zedoary, Calamus Aromaticus, each a drachm, red Roſes dry'd an ounce, Liquorice two ounces, Raiſins ſton'd half a pound, Cochineal, Saffron, each one drachm, beſt Brandy one gallon; infuſe for eight days, and ſtrain, to which add Loaf Sugar twelve ounces." [Compleat Body of Distilling]

1747 – In his *Pharmacopœia Universalis* James also mentions the plague connection, although his words suggest that not all physicians were in agreement over the worthiness of sundew: "*It grows in boggy Grounds, and flowers in* June *and* July. *The Virtues of this Plant are much controverted, ſome recommending it as good for the Phthiſis, and Plague, whilſt others, not without Reaſon, forbid the internal Uſe, on Account of its cauſtic Qualities.*" [Pharmacopœia Universalis]

1790 – By the time of William Meyrick's domestic *New Family Herbal* there seems to be no suggestion as to sundew being used for plague: "*Some authors very gravely tell us, that a water diſtilled from this plant is highly cordial and reſtorative; but it is more than probable that it never deſerved the character given of it in that reſpect. The leaves, bruiſed and applied to the ſkin, erode it, and bring on ſuch inflammations as are not eaſily removed. The ladies in ſome parts mix the juice with milk, ſo as to make an innocent and ſafe application for the removal of freckles, ſun-burn, and other diſcolourings of the ſkin. The juice, unmixed, will deſtroy warts and corns, if a little of it is frequently put upon them. Theſe are effects which pronounce its internal uſe dangerous...*" [New Family Herbal]

SWEET CICELY – *Myrrhis odorata*

The inclusion of this fragrant aromatic perennial species here is due to sweet cicely often being among herbalists' and plant lovers' favourite plants, more than for its small contribution towards any plague curing qualities. Sweet cicely does occasionally appear in plague remedies but in a minor way when compared to plants such as angelica, the two rue species, and tormentil.

These days sweet cicely is perhaps more likely to be grown in gardens than anywhere else, but in the wild it was a common plant in the northern parts of Britain. It belongs to the *Umbellifer* plant group and grows about two to four feet in height, having stout hollow stems, pale green leaves, and produces a straight aromatic fruit about an inch long that has a pleasant anise-like taste and aroma. The leaves too have an anise-like taste, but are also sweet and, along with the seeds, were once regular kitchen ingredients. On the medicinal front the plant was used herbally in digestive upsets, for coughs, as a tonic, and apparently in consumption.

In the wild sweet cicely is often found growing by waysides and hedgerows near old habitations, as well as dry hill pasture, although it can tolerate some moisture and varying degrees of light and shade.

1578 – The Dodoens-Lyte herbal appears to be referring to sweet cicely root when it says: "*They ſay alſo that it is good to be drōken in wine, in the time of Peſtilence, and that ſuche as haue dronken three or foure times of the ſame wine, ſhall not be infected with the plague.*" [New Herbal]

1583 – In Royet's French plague treatise, under a chapter entitled 'Remedes, deffenſifz, & preſeruatifz', there is the following extended text extract that reveals use of sweet cicely among a much wider group of plants, some of which the reader will already be familiar with. The text suggests a preventative herbal regime that changes each day, with consumption of a sweet cicely wine on day four: "*Quand le tēps ſera froid faut prēdre vn petit morceau de pain roſti trempé dans quelque bon vin odoriferant: Vn autre iour lon mangera vne figue auec la moitié d'vne noix qui ſoit bonne, & non point rance, ou moyſie auec quatre ou cinq feuilles de rue, & vn petit de ſel. Vn autre iour on boira de la poudre de l'herbe hypericon, autrement dicte Mille pertuis, le poids d'vn eſcu, eſtant diſſoute au ſoleil en vin, & eau de bugloſſe. Vn autre iour boire en vin d'vne autre herbe nommée mirrhis. Vn autre iour maſcher & aualler des grains de geneure, & de la veruaine, vn autre iour de l'angelique. Vn autre, de la Zedoaire. Vn autre, qui eſt le meilleur, du ſcordium. Vn autre, boire du vin ou il y ait trempé des doux de giroffle.*" [Excellent Traicte de la Peste]

1585 – Over in Italy sweet cicely also appears to have been recognized for its plague curing properties, Castore Durante's herbal recommending the root in wine for the plague: "*Dicono alcuni, che beuendoſi ogni dì due o tre uolte queſta radice nel uino, è ſalutifera nelle peſtilēza, e preferua da quella chi ſe la beue.*" [Herbari Nuovo]

1629 – Parkinson's gardening work *Paradisus Terrestris* has the following details regarding sweet cicely, and possibly households with garden space may well have grown the plant among their other general herbal medicine plants, since the 1625 London plague outbreak would have been recent history for Londoners: "*Sweete Cheruill, or as ſome call it, Sweete Cis, is ſo like in taſte vnto Aniſe ſeeds, that it much delighteth the taſte among other herbes in a Sallet: the ſeede is long, thicke, blacke, and cornered, and muſt be ſown in the end of Autumn, that it may lye in the ground all the Winter, and then it will ſhoote out in the Spring, or elſe if it be ſowne in the Spring, it will not ſpring vp that yeare vntill the next: the leaues are vſed among other herbes: the rootes likewiſe are not onely cordiall, but alſo held to be preſeruatiue againſt the Plague, either greene, dryed, or preſerued with ſugar.*" [Paradisus Terrestris]

1710 – From Salmon's following comments in *Botanolgia* it sounds as if he had no practical experience of the plant in plague, and quotes the Parkinson reference above regarding the root: "*The great sweet chervil, sweet cicely, called cerefolium magnum, ſive myrrhis... Their taſte is not unpleaſant (for which reaſon many put them into Sallets) and is not much differing from the Taſte of* Aniſeeds..." Salmon then goes on to make recommendations on the herbal use of the plant, although in the case of the distilled water it is rather unclear which part of the plant is used, though it follows on directly from details about the green seed:

"Diſtilled Water. *Being drank to four or ſix ounces, it is good to diſſolve congealed Blood, and provoke Urine.*

The Candied Roots. *They are very good to warm, comfort, and ſtrengthen a cold and weak Stomach, and excite* Venus: *and* Parkinſon *ſays, they are thought to be a good Preſervative in the time of Plague.*" [Botanologia]

TORMENTIL – *Potentilla erecta*

Tormentil was highly regarded for its astringent and vulnerary qualities in the past and appears quite regularly in old plague remedies. It is a species that would have been very familiar to our ancestors, where tormentil would have appeared on commons and heaths in association with species such as broom, gorse, tussock grass, woodrush and bilberry for example. This yellow-flowered perennial prefers sandy type soils, but also stony habitats, and ranges between six and eighteen inches in height.

The reddish woody root of tormentil was a particular herbal favourite for treating dysentery and one of the many common names was Blood-root. The plant was used too in cases of ague, toothache, and as a gargle and lotion for mouth ulcers. Among the plethora of other local names were also Ewe Daisy, Five-fingers, Flesh-and-blood, Sept-foil, Shepherd's Knot, Shepherd's Root, Thormantle, and Turnmentille. The root, incidentally, offers up a red dye and may also be used for tanning leather.

Perhaps one of the best descriptions of the root's herbal qualities can be found in the *New Dispensatory* [1753] which reads: "*This root has an auſtere ſtyptic taſte, accompanied with a kind of aromatic flavour: it is one of the moſt agreeable and efficacious of the vegetable aſtringents, and is employed with good ſucceſs in all caſes where medicines of this claſs are proper. A tincture made from it with rectified ſpirit, poſſeſſes the whole aſtringency and flavour of the root, and loſes nothing of either in inſpiſſation; whilſt aqueous liquors elevate the whole of the aromatic part: the diſtilled water ſmells agreeably, ſomewhat like roſes.*"

1527 – Brunschweig's *Vertuous Book of Distillation* (the original dating back to 1500) has the following recipe and background instructions for a '*Water of Tormentylla*':

"*Consolida rubea in Latin. The beſt parte and tyme of his dyſtyllacyon is / the herbe and the rote with all his ſubſtaunce dyſtylled betwene bothe our lady dayes.*

It dronke in the mornig [sic] *faſtyng is good for the peſtylēce / & is a preſeruatiuum / that is a defending of the ſayd ſekenes the for peſtylence that ſhe can not come on a bodye.*

And yf a body hath the ſame ſekenes than he ſhall lete blode on the ſame membre / as it is ryghtfull / and after the lettynge blode / hym ſhall be gyuen two ounces of the ſame water / myxed with a dragma of Venys tryacle / and halfe an ounce and a quarter of an ounce of vynegre. Than [sic] *he ſhall be layde downe / and rubbed his handes and fete / with vynegre / herbe of grace / wormwode / & with ſalte / and than let hym well ſwete / and the other day doynge it alſo / thā he becometh hole agayne.*" [Vertuous Book of Distillation]

1578 – The Dodoens-Lyte herbal refers to tormentil thus: "*The leaues of Tormentill with their roote boyled in wine, or the iuyce thereof dronken prouoketh ſweate, and by that meanes it driueth out all venim from the harte: moreouer they are very good to be eaten or dronken againſt all poyſon, and againſt the plague or peſtilence. The ſame vertue hath the dryed rootes, to be made in pouder and dronken in wine.*" [New Herbal]

1579 – In Bullein's *Bulwarke of Defence*, which uses a 'conversation' between various characters to explain information, the text refers to properties attributed to tormentil by old medical writers then reviews more recent attribution: "*The new wryters ſay, it wyll cloſe & heale new woūds, and cleanſe the eyes. The pouder of the roote, wyth the iuice of Planteine, is drunke againſt the ſtopping or ſcaldyng of bryne, it is good againſt the poiſō. of peſtylence, and bloody flixe, kynges euyll, fylth in the mouth, and foreness or ſwelling of the ſtomache or Spleane, Liuer and Belly, and ſtoppeth bloud, if it be dronke in the decoction, the water, or pouder.*" [Bulwarke]

1636 – Stephen Bradwell's *Physick for the Sicknesse, Commonly called the Plague* written a decade after the 1625 London plague outbreak describes the following antidote: "*For young* Children, *there is nothing better or fitter then* Bole Armoniack, *or* Terra Lemnia, *with a little* Tormentill *roote, or* Citron pills, *made into fine Powder, and mixed with their meates, butter, and broths; for their breakfaſts.*" [Physick for the Sicknesse]

1640 – Around the same period London apothecary John Parkinson's *Theatrum Botanicum* tells its readers: "*... the juyce of the herbe or roote taken in drinke, not only reſiſteth all poyſon or venome of any creature, but of the plague and peſtilence it ſelfe, and peſtilential feavers, and infectious diſeaſes, as the pockes, meaſells, purples, &c. by expelling the venome and infection from the heart by ſweating: if the greene roote, is not at hand, or not to be had readily, the powder of the drie roote is as effectuall, to the purpoſes aforeſaid, to take a dramme thereof every morning; the decoction likewiſe of the herbes and rootes made in wine, and drinke, worketh the ſame effect, and ſo doth alſo the diſtilled water of the herbe and roote, rightly made and prepared, which is to ſteepe them in wine for a night, and then diſtilled in Balneo mariæ; this water in this manner prepared taken with ſome Venice Treakle, and there-upon being preſently laid to ſweate, will certainly by Gods helpe expell any venome or poyſon, or the plague, or any fever or horror, or ſhaking fit that happeneth, it is ſo effectuall in the operation againſt the plague...*" [Theatrum Botanicum]

1665 – Thomas Cocke's self-promoting plague pamphlet *Advice for the Poor by way of Cure & Caution* during the Great Plague of London describes some pills that the author personally used, according to Cocke: "*Inwardly I uſe this* (*) *Lozange, which being diſſolved in the Mouth and let down into the Stomack (by a Stiptick property, it hath to cloſe the Orifice of the Stomach and paſſages that lead to it) they reſiſt the Attraction of Malignancy into the inward parts, by ſtrengthning* [sic] *the*

Lungs, and wonderfully affifting the Heart (by fixing Humours) to refift fudden Death: The ufe of which are of very great benefit for Lawyers, Clergy-men, and Citizens who have publique converfe and concernments with people: I take no other my felf, who am not ignorant of moft that are Extant.

(*) *Take of the true prepared* Orientall Bole Florentine orrice, *of each halfe a pound,* Sugar *two pound,* Tormentil *roots one Ounce,* Mirrh *in powder half an Ounce, with* Gum Dragon *diffolved in Vinegar, make a paft for* Lozanges..." [Advice for the Poor]

1665 – Nathaniel Hodges' personal account of the London plague published in the 1720 book *Loimologia* refers to a Plague Water from the 'College' which he says was effectual, but that it would need adjusting to the constitution of the patient: "*Rx. Radic. tormentillae, angelicae, paeoniae, zedoariae, glycirrhizae, helenii ana ʒ ẞ. fol. Salviae, Chelidoniae, rutae, fummitat: rorifmarini, abfynthii, roris folis, artemifiae, pimpinellae, dracunculi, fcabiofae, agrimoniae, meliffae, cardui, betonicae, centaurii min. fol. & flor. calendulae and* M i. *(alii addunt flor. papaveris errat: paralyf. ana p. iij.) incifa, & contufa infundantur per triduum in lib. viij. vin. alb. opt. dein F. cauta diftillation & liquor ufui refervetur...*" [Loimologia]

1703 – Circulating the English professional medical establishment during the 17th-18th century divide was the following medical wisdom from Germany in *Etmullerus Abridg'd*: "*A Loofenefs or Gripings of the Gutts attending a Plague, are accounted for, by exhibiting Diafcordium, Opium, abforbant Powder, Extract of Treacle, or that of Tormentil, Camphyr, Vinegar, and dulcified Spirit of Salt.*" [Etmullerus Abridg'd]

1710 – Salmon's *Botanologia* describes quite an extensive number of tormentil extracts for plague use, although a decoction in wine was not used and the spiritous tincture only sometimes. The main key tormentil extracts for use were:

"The Liquid Juice of the whole Plant. *It ftops all Fluxes of Blood or Humors in Man or Woman... refifts all Poifon, and the Plague and Peftilence it felf, and all Peftilential Difeafes... Dofe 3 or 4 Spoonfuls Morning and Night, in a Glafs of Mull'd Sack, or other Stiptick or fit Wine.*

The Effence. *It has all the former Virtues with advantage, as being the more efficacious Medicament... It is an effectual Antidote or Counter-Poifon, Antipeftilential, and a Cure for the yellow Jaundice... Dofe 3 or 4 Spoonfuls Morning, Noon and Night, in fome proper Wine or Liquor.*

The Diftilled Water... If it is given with 2 Drams of Venice Treacle or Mithridate *diffolved in it, and the Patient put to Sweat... it will potently provoke Sweat, and fo expel Poifon, and defend the Heart and Vitals from the Malignity and Infection of the Plague and Peftilence, and from the danger of any Peftilential or Infectious Difeafe.*" [Botanologia]

VIOLET – *Viola odorata*

This author has never come across sweet violet in anything but a domestic setting, where it usually finds favour among plant growers but often escapes into the immediate hinterland around human habitation. The plant does appear in the wild, perhaps more at home in humus-rich soils of shaded woodlands than elsewhere but does tolerate sunlight, and has a creeping, running, root system.

Perennial sweet violet rarely grows more than six inches tall and produces its' sweet smelling violet-white flowers from around March through May. These were used by the perfumery business, though without doubt synthetic substitutes now exist. Herbally, the plant was used for a variety of complaints, various parts having laxative, emetic, diuretic, expectorant, and hypotensive qualities.

1578 – In his *Dialogue* Bullein seems to draw down on the wisdom of Galenic thought for this recipe: "*The sirrupes of Violettes, of Sorrell, of Endiue, of sower Limondes of eche like, mingled with Burrage water, a Ptisane made of Barlie mingled together, is verie holsome to drinke: put in the pouder of bole Armoniack, whiche is of a singular vertue to coole; for Galen did help thousands at Rome with the same Bole and the* Theriaca *mingled together, in a greate pestilence.*" [Dialogue Against the Fever Pestilence]

1590 – Under a section on the Pestilence in Philip Barrough's *Method of Physicke* there is a thirst-quenching potion that mixes violet julep with sorrel and citrus ingredients among others. Syrup of violets pops up quite frequently in plague-related recipes, very often as an admixture that provides the potion some fluid quality: "*If the patient have vehement thirſt, he may vſe this potion. Rx. Iulep of the Violets ʒ.iiij. ſyrupe of the ſharpe iuyce of Cytrons ʒ.j.ß. ſyrupe of ſower Endive ʒ.ij. of the decoction of ſorrell, ſcabious, and floures of bugloſſe. ʒ tenne, or ſo much of their diſtilled water wherein barley hath been ſodden a little, and commixe with it iuyce of roſes, or ſorrell, or lymons, or of unripe grapes, and miniſter it in ſteede of drinke.*" [Method of Physicke]

1615 – Under sudorifics Canadelle has this potion of violet flowers and mellisa (lemon balm probably): "*Prens eau de meliſſe, de bourrach, & de fleurs de violettes, de chacune deux onces: eau roſe & de ozeille, de chacune vne once et demie: incorpore les au mortier de marbre, auec confection alchermes, & perles preparees, de chacune trois drachmes, y adiouſtant ſur la fin eau de canelle trois demi once.*" [Traicte et Familier de la Peste]

1665 – Lovell's *Panbotanologia* provides the following background to a syrup of sweet dog violet leaves that had some place in plague remedies: "*The* ſyrup *alſo helpeth the inflammation of the lungs and breaſt, the pleuriſe, cough, and feavers and agues in children, 8, or 9. drops of oile of vitrioll being mixed with* unc. l. *of ſyrrup* and a *fpoonful* give at once: *it alfo help burning feavers, and peftilent difeafes, all inflammations of the throat, mouth, uvula fquinancy and the epilepfie in children.* [Panbotanologia]

1710 – At the start of the following century Salmon presents two syrups of violet flowers in his *Botanologia*. In the first the juice is mixed with sugar then heated, while the second is an aqueous version. According to Salmon the syrup: "*... cools, moistens, allays the heat of fevers, comforts the stomach, chears the heart, and resists putrefaction.*"

The instructions read: "*With an Infufion of Water. Take frefh Flowers of Violets a Pound, fair Water boiling hot a Quart; ftop them clofe up in a Glafs Matrafs or* Vefica *for a day, then ftrain out by preffing; in the ftrained Liquor two Pounds, diffolve of Double Refined Sugar, four Pounds, by the heat of a Bath, and taking off the Scum, make it into a Syrup without boiling.* It has all the former Virtues, but is lefs powerful, and therefore may be given in double the quantity. Either of thefe Syrups will be much more effectual in hot, burning, malign and peftilential Fevers, if they be made a little Acid (when given) with fome few Drops of the *Spirits or Oils of Sulphur or Vitriol...*" [Botanologia]

1790 – By the end of the century sweet violet gets no mention in Meyrick's *New Family Herbal* in relation to violets being used in plague. The author simply tells his domestic readership: "*The flowers are cooling, emmollient, and gently purgative, but they lofe the greateft part of thefe virtues in drying, and as they can only be had frefh in the fpring, the beft method of ufing them is in form of a fyrup, which, when carefully made, is very pleafant, and contains all the virtues of the flower. It is excellent, mixed with a fmall quantity of oil, to keep the bowels of children gently open, and may likewife be given with great fuccefs againft habitual coftivenefs in grown people; it is alfo good in coughs, hoarfeneffes, and other diforders of the breaft. The feeds, dried and powdered, work gently by ftool, increafe the urinary difcharge, and are excellent in the gravel, and all complaints of the kidneys and bladder. The leaves are emmollient, a decoction of them is frequently an ingredient in glyfters, for foftening and lubricating the bowels.*" [New Family Herbal]

WALNUT, ENGLISH – *Juglans regia*

Most readers of this book will probably be familiar with the walnut's fruit, the oil for culinary use, or the beautiful colour and structure of the wood used in furniture. The tree appears to have been introduced to Britain in the distant past from areas east of the Black Sea, and became widely cultivated for its nuts and wood. It prefers a good, rich, somewhat dry soil, and mainly grows in the southern parts of Britain, being less frequently found the further northwards one travels.

The dried inner bark produces rather severe vomiting and was used to clear out the stomach of slime and mucus, though was only recommended to persons of robust constitution. The external green shell of the fruiting part was a reasonably powerful astringent, the expressed juice being commended as a gargle and for mouth and throat ulcers. The outer husk and shell, and peel of the kernels, was an esteemed sudorific. Boiled with sasparilla and guaiacum wood the decoction was used for treating rheumatism, venereal complaints and expelling worms. An infusion of the nut shells, it was said, could even destroy tapeworms. Infusions of the large, reddish-tinged, aromatic, leaves were also said to have the same worming property, while the expressed oil of the nut was also claimed to achieve the same. In terms of the plague *Juglans* regularly appears in treatment remedies.

1527 – In the *Vertuous Book of Distillation* (originally from 1500) Brunschweig has a recipe for a '*Water of grene walnuttes*' with the following instructions for its use in times of the plague: "*Two or thre tymes in a daye dronke of the fame / at eche tyme an ounce / or an ounce and a halfe / is very good agaynſt all hete / and clowtes wet in the fame and layd ther on It is alfo good for the blacke blaynes / and for the blaynes name Anthrax / and they be the blaynes of the peſtylence / lynen clowtes or towe wet in the fame water and layd ther upon / two or thre tymes in a daye... Dronke of the fame water two oūces or two ounces and a halfe / is good agaynſt the peſtylence.*" [Vertuous Book of Distillation]

1541 – Thomas Elyot's *Castel of Helth* mixes walnuts with that other anti-plague stalwart, rue. The following recipe orginates with Galen and Apollonius from a much earlier time, and can be seen pervading plague texts of the 16th-17th century period: "*Of two drye nuttes, as manny fygges, and .xx. leaues of Rewe, with a grayn of ſalt, is made a medicine, wherof if one doo eate faſtyng, nothinge which is venomous, may that day hurte them, and it alfo preferueth agaynſt the peſtilence, and this is the very ryght Mithridate.*" [Castel of Helth]

1557 – The French *Remedes Secretz* of Euonyme Philiatre appears to add treatment of plague sores to walnut's uses: "*L'eau de noix iuglandes (ce font communes de noyer) non encore meures, apreſtee enuiron la feſte fainct Iean: & appliquee par dehors, eſt bonne es playes & vlceres chaux, & au charbon peſtifere. Semblablemēt beüe à la qualité de deux ou trois onces, elle refreſchit & reſiſte à la peſtilence."* [Tresor des Remedes Secretz]

1578 – In the Bullein plague treatise, *Dialogue Against the Fever Pestilence*, the suggestion appears to be that an electuary containing walnut should always be available during plague times: *"Be neuer without the electuarie of nuttes, thus made: cleane Whalnuttes xx, fatte Figges xiij, herbe Grace two handfull, of Worme wodde, Fetherfu, or rather Cotula Fætida, called* Buphthalmus, *called Oxe eye, and Scabios, of eche one handfull, the rootes of* Aristolochia longa *halfe an vnce,* Aristolochia rotunda *an vnce and a halfe, The rotes of Turmentill and of the lesser Burre called* Petasitus, *Pimpernell, of eche ij vnces and a halfe, the leaues of the berie Dictamini one handful, Bay beries iiij Dragmes, the pouder of Hartes horne twoo* [sic] *drames and a halfe, Maces, Myrrhe* [sic]*, Bole Armoniacke, and the yearth of Limondes, of eche Dragmes three, Salt of the Sea a dragme and a halfe,* Nux vomica *dragmes twoo, Buglos flowers one handfull, stamped together by arte & with clarified honie make it; this is good to be eaten a dragme euerie mornyng."* [Dialogue Against the Fever Pestilence]

1666 – From the collected plague recipes of Thomas Willis in *Plain & Easie Method* [1691], we find one plague remedy expressed as being: *"For the Poorer Sort, that Medicine of the Ancients, may be proper..."* The recipe that follows has a familiar make-up: *"Take of* Rue *two handfuls, Figs and* Walnut-Kernels, *of each twenty four, common Salt half an Ounce; Which beat all together in a Mortar, till it be well mix'd; Take of it as much as a Nutmeg every Morning and Night."* [Plain & Easie Method]

1671 – In the French work *Recueil des Remedes et Secrets*, that draws its knowledge-base from English medicine of the period, and Kenelm Digby's work, there is a long anti-plague remedy attributed to Mayern, a well-known physician in English medicine. Among the ingredients are some familiar faces, blessed thistle, marigold, angelica and water germander: *"Il faut prendre de noix vertes & les piler dans un mortier, puis en tirer le jus par expreſſion, puis prenez jus de Baume, jus de Chardons benits, jus de Calendula de chacun trois chopines, racines de Lapatum, racines d'Angelique, de chacune demy livre: Genetta la plante entière; c'eſt à dire, l'herbe & racine, douze onces: les feüilles de Scordium, deux poignées: du Theriaque de Veniſe & du Mitridat, de chacun quatre onces: jus de Citrons chopine, vin de Canarie trois chopines: du Safran demy dragme. Digerez, cela tout enſemble dans une cucurbite l'eſpace de deux jours, puis le diſtilez, & quād vous en aurez tiré la moitié, faites paſſer par un linge ce qui reſte dans la cucurbite, puis le diſtilez juſq'à ce qu'il loit en conſiſtance de miel; Vous le mettrez dans un pot de Fayance, pour vous en ſervir dans le temps de la contagion, avec l'eau diſtillée."* [Recueil des Remedes et Secrets]

1710 – For his domestic readership Salmon gives his audience in *The Family Dictionary* the following remedy for plague sickness: *"Take Walnuts when the Green Husk is on them, and before the Shell is hardned* [sic] *underneath; put them when bruifed, to fteep in White Wine eight days; then with fome Bawm* [sic]*, Rue, and tops of Fetherfew, and Wormwood a little bruifed, put them into an Alembick, and Diftil them. You may drink an ounce and a half of the Water, Morning, Noon and Night, put into it fome perfumed Comfits, and ftir them well about till they are diffolved."* [The Family Dictionary]

1720 – This 18th century reprint of Culpeper's *Pharmacopœia Londinensis* – the original work being published in the previous century – provides a recipe for treacle water. This writer has not been able to track down an original first edition of the book so is uncertain whether there have been any amendments to the original recipe composition or Culpeper's personal comments. While walnut juice may have been a key ingredient here, there are several other herbally trusted anti-plague plants too, angelica, butterbur, masterwort, rue and scordium:

"Aqua Juglandum compositae. Or, Walnut Water Compound

College. Take of the juice of Green Walnuts four pound, the juice of Rue three pound, juice of Carduus Marigolds and Balm, of each two pounds, green Petafitis Roots one pound and an half, the Roots of Burs one pound, Angelica and Mafterwort, of each half a pound; the leaves of Scordium four handfuls, old Venice Treacle, Methridate [sic]*, of each eight ounces; Canary Wine twelve pound, Vinegar fix pound, juice of Lemmons two pound; digeft them two days; either in Horfe dung, or in a Bath, the veffel being, clofe fhut then diftil them in fand, and the diftillation you may make a Theriacal Extraction.*

Culpeper. *This water is exceeding good in all Fevers, efpecially Peftilential; it expelleth venemous Humours by fweat. It Strengthens the Heart and Vitals. It is an admirable Counter poyfon; fpecial good for fuch as have the Plague, or are poyfoned or bitten by venomous beafts..."* [Pharmacopœia Londinensis]

WOOD SORREL – *Oxalis acetosella*

Readers familiar with the countryside may know diminutive wood sorrel, the delicate, shade-loving, plant with pleasantly acid tasting leaves. Wood sorrel prefers habitats with little competition from larger plants, and is frequently found on sloping hedgerow banks or the floor of woodland where it generally blooms around April and May.

Wood Sorrel was known by many names, the most usual in materia medica and herbal remedy books being *Alleluia* and *Lujula*. However, the plant also had a variety of quaint local names too; Bread-and-Cheese, Bird's Clover, Sour Clover, Cuckoo-spice, Cuckoo's Victuals, Sour Grass, Green Sauce, Hallelujah, Hare's Meat, Lady's Clover, Sleeping Beauty, Sleeping Clover, Stabwort, Wood-sour and Woodsower. The name Stabwort reflects the use of the plant for wounds, stabs, and skin punctures, at least that was the theory. The plant was believed to be endowed with cooling, antiscorbutic, and diuretic properties, with an infusion given in cases of fever. *The New Dispensatory* [1753] explained the herbal qualities of the plant this way: "*In taſte and medical qualities, it is ſimilar to the common ſorrel but conſiderably more grateful, and hence is preferred by the college. Boiled with milk, it forms an agreeable whey; and beat with ſugar, a very elegant conſerve, which has been for ſome time kept in the ſhops, and is now received in the diſpenſatory.*"

1623 – In the French plague text *Advis Sur la Nature de la Peste*, which was written at the time Cardinal Richlieu held sway in French politics, there is the following recipe that incorporates the leaves of sorrel (*ozeille*) as well as those of wood sorrel (*treffle aceteux*), with a dose being taken two hours before a meal: "*Les feüilles de ruë, de myrrhis, d'hypericum ou millepertuis, de veruene, ozeille, quinte-feuille, pimpinelle, treffle aceteux, ou aleluya, ſcabieuſe, ſcordium, ſoucy & galega, ou ruta capraria, ſont fort ſingulieres en decoction priſes au matin, tantoſt de l'vne, tancoſt de* l'autre, ſelon celles qu'on a plus à commandement, au poids de trois onces, ou enuiron, y adiouſtant vne once de ſyrop de limons, & ce deux heures auant le repas.*" [Advis Sur la Nature de la Peste]

1665 – During the Great Plague of London the Royal College of Physicians issued the pamphlet *Certain Necessary Directions* for the general public, or at least those who could read. Below are a 'preservative' electuary and another specified for pregnant women and children:

"*Take of the Conſerve of Wood-ſorrel two ounces, of Diaſcordium two drams, of the flour of Brimſtone very finely ground one dram, of Saffron three grains, of Syrup of Wood-ſorrel as much as is ſufficient to make an Electuary: for prevention, take a dram every morning faſting, during the imminent danger...*

For Women with child, Children, and ſuch as cannot take bitter things, uſe this.

Take Conferve of Red-Rofes, Conferve of Wood-Sorrel, of each two ounces, Conferves of Borage, of Sage-flowers, of each fix drams, Bole-Armoniack, fhavings of Harts-horn, Sorrel-feeds, of each two drams, Saffron one fcruple, Syrupe of Wood-Sorrel, enough to make it a moift Electuary; mix them well, take fo much as a Chefnut at a time, once or twice a day, as you fhall find caufe." [Certain Necessary Directions]

1665 – Thomas Cocke's pamphlet, *Advice for the Poor,* was another of the plague advice publications that appeared during the great London epidemic. Among the preventative remedies was the following, that provided the option of a cheaper mixture for the poor: *"One pound of the Conferve of Wood-Sorrel, of* London Treacle, *and* Bole Armoniack; *Sirup of Vinegar, or Sirup of* Citron, *of each one Ounce, is undoubtedly a far more fafe and Sovereign* Preventive, *taken morning and night, the quantity of a large Nutmeg. The poorer fort may put the* Bole *and* London Treacle *in a pint of Vinegar, and take a Spoonfull Morning and Night, firft fhaking it."* [Advice for the Poor]

1665 – A similar recipe to the one above is given in the 1720 reprint of Hodge's historical account of the Great London plague, *Loimologia.* The ingredient list for this *Electuary for the Poor* changes slightly, with the addition of rue (galegae) that familiar and trusted anti-pestilential herbal plant of the past: "Rx. *Conferv. lujulae, galegae ana* lib. fs. *calendulae* lib. j. *Theriac. Londin.* ʒ iij. *boli armen. vitriol.* ʒ iv. *cum fyr. limonum q. f. conficiatur Elect. Dof. ad* 3 ij. *vel* iij." [Loimologia]

1666 – Among Thomas Willis' collected remedies, published posthumously in *Plain & Easie Method* [1691], there is another 'preservative' that the text advises should: *"...be taken of every Morning, and again at Night, by thofe that live in infected places":* Take of Conferve of Wood-Sorrel *four Ounces;* Confectcio Liberans, *and* Mith-ridate, *of each half an Ounce; Salt of* Wormwood *two Drams; Confection of* Hyacinth *one Dram;* Tormentil Roots, *and fine* Bole, *of each half a Dram; Peftilential* Vinegar *half an Ounce; mix all with Syrup of* Citron; *Take as much as a Nutmeg, Night and Morning."* [Plain & Easie Method]

1666 – Also from the battleground of the Great Plague are the following two juleps from London apothecary William Boghurst:

"If the Party bee troubled with constant vomiting you may make such a Julep in stead of beer, and in this disease nothing is better to deny themselves much drinke, or drinke small draughts of this Julep.
Rx. Aq. lujulae, Citri totius ana ʒ *iiii aq. Cinnamoni hordeati, menthae, rosar. rub., ana* ʒ *s.s. succi aurantior civil,* ʒ *ii Syrupi e succo Cydonior* ʒ *ii s.s. ol. vitrioli* gt *xii m. capiat* C *ii quâque hora. If they bee disposed to bee loose which is dangerous such Julep as this will doe well:*
Rx. Rad. tormentillae, acetosae, Symphiti, Scorzonerae, ana ʒ *i fol lujulae, centinodii ana* M *i, rosar. rub.* M. *ii cornu cervini, eboris, ana* ʒ *i Cinnamoni* ʒ *s.s. Coq. aq. fervent et vino albo ana* C *s.s. ad* lb. *i s.s. Colaturae adde Syrupi de rosis siccis, Syr.*

myrtini ana ℥ ii *ol. vitrioli* gt xx m. *Cap. cochlearia sex ter quaterve in die."*
[Loimographia]

1710 – Salmon's professionally leaning *Botanologia* has little to say regarding the qualities of wood sorrel for plague use in the following century, simply advising: "The Juice … *cools Inflammations, takes away all preternatural heats, whether in the Stomach, Bowels, or habit of the Body, reſiſts putridity, and is moſt excellent againſt any Contagious ſickneſs, or Peſtilential Fever. Mixt with a ſir quantity of double refined Sugar, it makes a moſt incomparable Green Sawce."* [Botanologia]

1713 – In the re-published work of George Bate's, *Pharmacopœia Bateana*, Salmon provides Bate's original recipe for *'Syrupus Allelujae*, Syrup of Wood Sorrel' but offers very little extra personal advice in his follow-up commentary:

"Bate.] Rx. *Juice of Wood-Sorrel* ℔iiſs, *Red Roſe-water* ℔ſs. *Trebble* [sic] *refined Sugar* ℔iſs. *mix and digeſt in B.M. for long till the feces* [sic] *ceaſe to ſubſide.*
Salmon.] It is a cooling Cordial, and may be given in the height of any burning malignant or Peſtilential Fever…" [Pharmacopœia Bateana]

1751 – John Hill's professional *History of the Materia Medica* makes only a passing reference to wood sorrel being employed in fevers, with no specific mention of its plague applications: *"It is a very grateful Acid, and in Fevers quenches Thirſt, and takes off the Heat of the Stomach. It is recommended in Fevers of all Kinds, and in the Scurvy, and alſo in Obſtructions of the Liver and other* Viſcera. *It is ſometimes given in Decoction in Fevers, and the expreſſed Juice is mixed with the Juices of the other antiſcorbutic Plants, when intended againſt the Scurvy. Externally it is much recommended againſt inflammatory Eruptions of all Kinds in Decoction, which is to be uſed by Way of Fomentation. There uſed to be a Syrup and a diſtilled Water of* Lujula *kept in the Shops, but at preſent they are rejected, and only the Conſerve is retained."* [History of the Materia Medica]

1852 – Certainly by the time of the Thompson-Garrod *London Dispensatory*, which is a professional medical work, it appears that English physicians had moved onwards to citrus fruits for their fever remedies, with not a hint of plague is mentioned in the following details about wood sorrel: *"Wood-sorrell is refrigerant and antiseptic. Boiled in milk it forms a pleasant whey, which may prove a useful refrigerant in fevers, as may also the expressed juice, or the quadroxalate obtained from it diluted with water: but although they are much extolled in inflammatory, bilious, and putrid cases, by the Continental physicians, yet their place is well and easily supplied by lemon-juice, or the citric acid, dissolved in water. The recent herb, eaten as a salad, may be serviceable in scorbutic affections."* [London Dispensatory]

BIBLIOGRAPHY

Key to the structure of this book has been the chronological use of the herbal plants in the plague as well as medical thinking on the subject over the centuries, hopefully allowing better chronological analysis of the data. For this reason the bibliography is laid out in date order, with the first or prior publication date of an original work used as the chronological benchmark, not the re-published date of an original work. This format only concerns a handful of texts, mainly those of Ambroise Paré, originally published in the latter half of the 16th century, the remedies of Thomas Willis, Nathaniel Hodges, William Bullein, and a few others. The re-published date of the orginal work is given in brackets after the publisher's name. The chronological text extracts in the Plants section have the source noted in square brackets at the end, which hopefully will make it easier to quick-scan the pages for certain references.

14th c.   Opuscule relatif à la peste de 1348, composé par un contemporain, E. Littré. Bibliothèque de L'École Des Chartes. Pub. Schneider & Langrand, Paris. (1840-1844)

1483   Regimen Contra Pestilentiam, Jacobi Johannes. Pub. Wm de Machlinia. *English.*

1485   A passing good Little Boke. Pub. William de Machlinia, London.

1485   Regimen Contra Pestilentiam, Benedictus Kanuti. *Latin.*

15th c.   De Pestilentia Liber, Joannis de Burgundia. BL Sloane MS. 3449

1527   The Vertuous Book Of Distillation, Hieronymous Brunschweig. Ptd. by Laurence Andrew.

1541   Castel of Helth, Sir Thomas Elyot. Pub. T. Berthelet, London.

1542   Compendyous Regiment or Dyetry of Health, Andrew Boorde.

1543   Nouveau Traicte, Intitule Lædifice ou bastiment des receptes. Pub. Iehan Real, Paris.

1545   Briefue institution pour préserver & guérir de la peste, Jehan Guido. Sold by Nicolas Buffet, Paris.

1545   Le Tresor et Remede de la Peste, Jean Tibault. Sold by Angelin Benoyt, Lyon.

1551   De la Maniere de Preserver de la Pestilence, Benoit Textor. Pub. Iean de Tournes & Guil. Gazeu, Lyon.

1557   Tresor des Remedes Secretz, Euonyme Philiatre. Pub. Antoine Vincent, Lyon.

1563   De Medendis Humani Corporis Malis Enchordion, Petri Bayri. Pub. Perna, Basel.

1565   Difesa Contra la Peste, Marcello Squarcialupi. Francesco Moscheni, Milan.

1566   A Dial For All Agues, John Jones. Ptd. by William Seres, London.

1566   Traicte de la Peste, François Valleriole. Pub. Antoine Gryphius, Lyon.

1569	The Gouernance and Preservation of them that Feare the Plage, John Vandernote. Ptd. by Wyllyam How for Abraham Veale, London.
1570	L'Agriculture et Maison Rustique, Charles Estienne. Pub. Jaques Du-Puys, Paris.
1575	The Workes of that Famous Chirurgion Ambroise Parey. Ptd. by Richard Cotes & Willi Du-Gard, Sold by John Clarke, London. (1649)
1576	Instruttione Sopra la Peste, Michele Mercanti. Ptd. Vincent Accalto, Rome.
1578	A Dialogue Against the Fever Pestilence, William Bullein. Pub. Early English Text Society / Trubner & Co., London. (1888)
1578	Niewe Herbal, Rambert Dodoens, trans. Henry Lyte. Pub. Gerard Dewes, London
1579	An Hospital for the Diseased, TC. For Edward White, London.
1579	Bulwarke of Defence against all Sickness, Soarnesse and Woundes, William Bullein. Ptd. Thomas Marshe, London.
1582	Epydiomachie ou Combat de la Peste. Pub Robert Coulombel, Paris.
1583	Excellent Traicte De La Peste, Antoine Royet. Pub. Iean Durant.
1585	Herbari Nuovo de Castore Durante, Iacomo Bericchia & I. Tornierii, Rome.
1587	Thesaurus Pharmaceuticus, Caspar Schvenckfelt. Ex Officina Frobeniana, Basel.
1590	The Method of Physicke, Philip Barrough. Ptd. Richard Field, London.
1593	A Defensative against the Plague, Simon Kellwaye. Ptd. by Iohn Windet, London.
1596	A Rich Store-House or Treasury for the Diseased, A.T. Ptd. for Thomas Purfoot and Raph Blower, London.
1602	Traicte de la Peste, Caesar Morin. François Iacqvin, Paris.
1614	Secrets du Seigneur Alexis Piemontois. Ptd. by Robert de Rowes, Rouen.
1614	Traicte de Peste, Jean Vigier. Pub. Jean Anth. Huguetan, Lyon.
1615	Histoire Generale des Plantes, Jaques Dalechamps & Jean des Moulins. Pub. Guillaume Rouille, Lyon.
1615	Traicte et Familier de la Peste, Moyse Canadelle. Pub. Estienne Gamonet, Geneva.
1617	The Surgions Mate, Iohn Woodall. Ptd. for Laurence Lisle, London.
1623	Advis Sur la Nature de la Peste, M. François Citoys. Pub. Sebastien Cramoisy, Paris.
1624	The Method of Physicke, 6th Ed., Philip Barrough. Ptd. Richard Field, London.
1625	An Excellent and best Approuued Treatise of the Plague, Thomas Thayre. Ptd. for Thomas Archer.
1629	Paradisi in Sole Paradisus Terrestris, John Parkinson. London.
1633	Storehouse of Physical & Philosophicall Secrets, Thomas Harper.
1636	Haven of Health, Thomas Coghan. Ptd. by Anne Griffin for Roger Ball, London.

1636	Physick for the Sicknesse, Commonly called the Plague, Stephen Bradwell. Ptd. for Beniamin Fisher, London.
1640	Theatrum Botanicum: The Theater of Plants, John Parkinson, Apothecary. Ptd. by Thomas Cotes, London.
1641	Charitable Pestmaster, Thomas Sherwood. Ptd. for John Francklin, London.
1653	Epitomie of most experienced, excellent and profitable secrets appertaining to physicke and chirurgery, Owen Wood. Ptd. by E.C. for Michael Spark, London.
1665	A Practical Method as Used for the Cure of the Plague, Sir Charles Scarborough. Pub. B. Lintot, London. (1722)
1665	Advice for the Poor by way of Cure & Caution, Thomas Cocke.
1665	Certain necessary Directions as well For the Cure of the Plague, Royal College of Physicians of London. Ptd. By John Bill, Christopher Barker, London.
1665	Loimologia: An Account of the 1665 London Plague, 2nd Ed., Nathaniel Hodges & John Quincy. E. Bell & J. Osborn, London. (1720)
1665	London's Deliverance predicted, John Gadbury. Printed by J.C. for E. Calvert, London.
1665	Panbotanologia, or Compleat Herball, Robert Lovell. Ptd. by W. H. for Ric. Davis, Oxford.
1666	Loimographia, William Boghurst, Joseph Payne, Ed. Pub. P. Shaw & Sons, London. (1894)
1666	Loimotomia, or the Pest Anatamized, George Thomson. Ptd. for Nath. Crouch, London.
1667	The London Distiller, J. French. Pub. Thomas Williams, London.
1668	The English Housewife (Way to Get Wealth), Gervaise Markham. Ptd. for George Sawbridge, London.
1670	L'Ordre Public pour la Ville de Lyon. Antoine Valançol, Lyon.
1671	Recueil des Remedes et Secrets, 2nd Ed., Jean Malbec de Trefel. Pub. Guillaume Henry, Liege.
1674	Medicine et Chirurgien Des Pauvres, 4th Ed., A Doctor. Pub. Edme Couterot, Paris.
1677	Anatomia Sambuci, or the Anatomy of Elder. Collected in Latine by Dr Martin Blochwich. Ptd. for H. Brome and Thomas Sawbridge.
1682	La Pratique de Medecine, Lazare Rivere. Pub. Jean Certe, Lyon.
1685	The London Practice of Physic, Thomas Willis. Ptd. for Thomas Basset, London.
1691	Plain & Easie Method, Thomas Willis. Ptd. for W. Crook, London.
1693	Pharmacologia, Seu Manuduction ad Materiam Medicam, Samuel Dale. Pub. Sam. Smith & Benj. Walford, London.
1696	La Medecine Aisée, Le Clerc. Pub. Estienne Michallet, Paris.
1703	Etmullerus Abridg'd: or, A Compleat System of the Theory and Practice of Physic, 2nd Ed. Ptd. for Andrew Bell and Richard Wellington, London.
1706	Les Admirables Secrets d'Albert le Grand. Cologne.

1707   Histoire des Plantes de L'Europe, Caspar Bauhin. Pub. Nicolas de Ville, Lyon.

1710   Botanologia, William Salmon. Ptd. for H. Rhodes and I. Taylor, London.

1710   The Family Dictionary: or, Houshold Companion, 4th Ed., William Salmon. Ptd. for N. Rhodes, London.

1712   Universal Library: Or Compleat Summary of Science, Vol. 1. Ptd. for George Sawbridge, at the Three Flower-de-Lys in Little Britain.

1713   Pharmacopœia Bateana, 4th Ed., William Salmon. Ptd. for William Innys, London.

1714   Collection of Above Three Hundred Receipts, By Several Hands. Ptd. for Richard Wilkin.

1720   Pharmacopoiea Londinensis, Nicolas Culpepper. Ptd. By John Allen for Nicholas Boone et al., London.

1720   Practical Treatise of the Plague, Joseph Browne. Ptd. for J. Wilcox, London.

1721   Avis de Precaution Contre la Maladie Contagieuse de Marseilles, 2nd Ed. Pierre Jos. Zappate, Turin.

1721   Discourse of the Plague, George Pye. Ptd. by J. Darby, and sold by J.Roberts, A.Dodd, London.

1721   Some Observations Concerning the Plague, Anon. Ptd. Dublin, and reprinted for J. Roberts, London.

1722   Discourse of the Plague, Sir Richard Blackmore. Ptd. for John Clarke, London.

1722   Nouvelle Reflexions sur L'origine, la Caufe &c de la Peste. Sr. Manget, MD. Pub, Philippe Planche, Geneva.

1722   Pharmacopœia Officinalis & Extemporanea, 4th Ed., John Quincy. Ptd. for E. Bell, et al. London.

1723   Febrifugum Magnum, 6th Ed., John Hancocke DD. Ptd. For R. Halsey.

1724   Medicina Britannica, 2nd Ed., Thomas Short. Ptd. for R. Manby and H. Shute Cox, London.

1727   Materia Medica, Edward Strother (Trans.), Paul Harman. Charles Rivington, London.

1731   A Compleat Body of Distilling, 2nd Ed., G. Smith. Pub. Henry Lintot, London.

1737   The Complete Family Piece, 2nd Ed. Ptd. for A. Bettesworth and C. Hitch, et al., London.

1739   Curious Herball, Elizabeth Blackwell. John Nourse, London.

1739   The Compleat Housewife, 9th Ed., Elizabeth Smith. J & J Pemberton, London.

1743   An Historical Account of the Most Remarkable Pestilential Distempers, R. Goodwin MD. Ptd. & Sold by Richard Burdekin, York. (1832)

1744   Discourse on the Plague, 9th Ed., Richard Mead. Ptd. for A. Millar & J. Brindley.

1750   A General System of Surgery, 4th Ed., Laurence Heister. Ptd. for W. Innys et al., London.

1751	History of Materia Medica, John Hill. Ptd. for T. Longman, C. Hitch and L. Hawes, London.
1753	The New Dispensatory. Ptd. for J. Nourse, London.
1758	Cook's Pocket Companion & Universal Physician, Lydia Honeywood. Ptd. for J. Staples, London.
1761	Theory and Practice of Chirurgical Pharmacy, Ptd. for J. Nourse, London.
1761	Primitive Physick, 9th Ed., John Wesley. Ptd. by W. Strahan, London.
1762	Primitive Physick, 10th Ed., John Wesley. Ptd. by William Pine, Bristol.
1770	Lectures on the Materia Medica, Chas. Alston, John Hope. Ptd. for E & C Dilly, London.
1770	Marine Practice of Physick & Surgery, Vol.2., William Northcote.
1790	New Family Herbal, William Meyrick. Ptd. Thomas Pearson, Birmingham. Sold R. Baldwin, London.
1791	An Account of the Principal Lazarettos in Europe, John Howard. Pub. Johnson, Dilly & Cadell, London.
1794	Every Man his Own Physician, 7th Ed., John Theobald. Ptd. by Jones, Hoff & Derrick, Philadelphia.
1799	An Account of the Plague Which Raged at Moscow, De Mertens. Pub. F&C Rivington, London.
1802	Le Medecin Herboriste by a Medical Botanist. Pub. Servière, Libraire, Paris.
1813	A Brief Description of the Plague, Richard Pearson. Pub. Thomas Underwood, London.
1819	The Treasure of Health, Lewis Merlin. Philadelphia.
1821	The History of the Plague in Malta etc., J.D. Tully. Pub. Longman, Hurst, Rees et al., London.
1822	Manuel Préservatif et Curatif de la Peste, Martin de Saint-Genis. Ptd. S. Darneau, Lyon.
1822	New British Domestic Herbal, John Augustine Waller. Ptd. for E. Cox & Son, London.
1823	The Universal Herbal, Thomas Green. Pub. Henry Fisher, London.
1833	The Black Death in the Fourteenth Century, I.F.C. Hecker, Trs. B.G. Babington. Pub. A. Schloss, London.
1835	Family Physician, Daniel Whitney. New York.
1839	The Principles and Practice of Medicine, John Elliotson. Joseph Butler, London.
1852	London Dispensatory 11th Ed., Anthony Thompson, Ed. Alfred Garrod. Pub. Longman, Brown, Green & Longmans, London.
1864	Medicines, Their Uses and Mode of Administration, 6th Ed., J. Moore Neligan, Rawdon Macnamara, Editor. Fannin, Maclachan & Longman, Dublin, Edinburgh and London.
1870	The Diary and Correspondence of Samuel Pepys, Braybrooke Edition. Pub. Alexander Murray, London.
1903	A Manual of Plague, William Jennings. Pub. Rebman Limited, London.

www.ingramcontent.com/pod-product-compliance
Lightning Source LLC
Chambersburg PA
CBHW071702200326

41519CB00012BA/2603